Performance Analysis of Cooperative Networking with Multi Channels

This book covers wireless cooperative communication and advanced communication techniques for research scholars and post-graduate students.

Features:

- This book will be the reference book for cooperative communication.
- It addresses the problems in small-scale cooperative communication.
- It presents cooperative routing algorithms for large-scale cooperative networks with the constraint of throughput and transmission time.
- It presents energy-efficient transmission approach by making use of multiple radio terminals.
- It presents adaptive routing algorithm for large-scale cooperative network under mobility environment.

Performance Analysis of Cooperative Networking with Multi Channels

Praveen Kumar Devulapalli,
Sushanth Babu Maganti, and
Pardhasaradhi Pokkunuri

CRC Press
Taylor & Francis Group
Boca Raton London New York

CRC Press is an imprint of the
Taylor & Francis Group, an **informa** business

Designed cover image: © rawpixel.com on Freepik

MATLAB® is a trademark of The MathWorks, Inc. and is used with permission. The MathWorks does not warrant the accuracy of the text or exercises in this book. This book's use or discussion of MATLAB® software or related products does not constitute endorsement or sponsorship by The MathWorks of a particular pedagogical approach or particular use of the MATLAB® software.

First edition published 2025
by CRC Press
2385 NW Executive Center Drive, Suite 320, Boca Raton FL 33431

and by CRC Press
4 Park Square, Milton Park, Abingdon, Oxon, OX14 4RN

CRC Press is an imprint of Taylor & Francis Group, LLC

© 2025 Praveen Kumar Devulapalli, Sushanth Babu Maganti, and Pardhasaradhi Pokkunuri

ISBN: 9781032713977 (hbk)
ISBN: 9781032713991 (pbk)
ISBN: 9781032714011 (ebk)

DOI: 10.1201/9781032714011

Typeset in Adobe Caslon
by Newgen Publishing UK

Contents

Preface

First, in Chapter 1, we discuss cooperative communication by highlighting wireless channel impairments and addressing cooperative architecture gains. We offer a systematic guide to understanding fundamental principles and implementing valuable taxonomies. The literature review supports the need for a thorough investigation into existing cooperative protocols, and the proposal for a design for developing a cooperative communication model is presented in Chapter 2. Furthermore, the central hypothesis presented in the literature review indicates that meaningful results may be obtained by efficiently selecting relays with correct channel estimation. The rest of this book, which is characterized by its focus on how and when to select relays, will seek to address the lack of studies on successful relay selection in the form of a cooperative communication model under such channel conditions.

In Chapter 3, two sophisticated routing algorithms based on MACC and distance parameters are implemented to maximize the benefits of CR in a multi-radio-multi-channel large-scale wireless network. The combined throughput and transmission time of the proposed algorithms are similar to the IACR algorithm. The best relay node with the best channel capability is identified among candidate relay nodes by using nodes near to LOS between source and destination and away from the source node as a guide. The dynamic CA with high channel functionality is formulated at the access point using the global chart. According to the simulation, the proposed cooperative relay selection algorithm reduces propagation delay and average hop count by 27.5 and 32.5 percent, respectively.

Chapter 4 proposes an energy-efficient transmission scheme for large-scale multi-radio multi-hop cell ad hoc networks to reduce energy usage. Better BER efficiency can be accomplished without increasing network power by introducing spatial multiplexing. The suggested EET scheme utilizes spatial multiplexing and dynamic power allocation to the nodes to achieve the target BER. The proposed scheme saves about 99 percent of energy as opposed to fixed power transmission, according to the simulation results. The efficiency of the proposed EET is often evaluated in terms of the number of hops for a normalized distance between the source and destination. According to the simulation, we will save up to 91 percent of energy while we use the EET.

A hybrid multi-hop cooperative routing algorithm for LC-MANET was described in this work. To counteract the mobility impact and reduce the total number of hops, combine clustering and location-based techniques. Incorporated optimization mechanisms in each hop and obtained the optimum number of cooperative nodes by maximizing the number of transmitters and receivers simultaneously. When opposed to the conventional routing approach, simulation results indicate that the proposed algorithm saves up to 53.42 percent in energy consumption.

Simulations were used to draw the results in Chapters 3, 4 and 5. Rayleigh fading was the channel model used in the simulations. This model, we assume, is a decent approximation for the channels. It will be more informative, however, to simulate the schemes and algorithms suggested in this research utilizing more practical channel models. However, it could be desirable to expand the resource distribution to involve resources, utilizing either short-term or long-term power restrictions. Similarly, by having many people in the framework to share system resources and addressing the resource allocation issue, the work may be extended. Furthermore, the selection schemes, unlike the optimum algorithm, are subject to node numbering. As a consequence, we propose the development of an efficient numbering scheme that enhances node-selection efficiency. In the future, the impact of imperfect synchronization on the Large-Scale Cooperative MANET model will be considered.

About the Authors

Praveen Kumar Devulapalli received his B.Tech, M.Tech in Electronics and Communication Engineering from Jawaharlal Nehru technological University, Hyderabad in 2006 and 2010 respectively. He was awarded a Doctorate degree in Electronics and Communication Engineering at Koneru Lakshmaiah Education Foundation, Vijayawada in 2021. He is presently working as Assistant Professor in Department of Electronics and Communication Engineering at Vardhaman College of Engineering, Hyderabad. He is Senior member of IEEE. He is a program chair for two International Conferences, ICETEMS and BROADNETS. He guided 8 PG and 23 UG projects. He has published more than 30 papers in international journals and conferences. His research interests are in the areas of wireless mobile communication, cellular networking, distributed cooperative communication, MIMO and signal processing applications.

Sushanth Babu Maganti received his B.E. in Electronics and Communication Engineering in 2002 from North Maharastra University and his M.Tech. degree from Jawaharlal Nehru Technological University, Hyderabad in 2008. He was awarded Doctorate degree in Wireless Communications at Jawaharlal

Nehru Technological University, Hyderabad in 2014. He is presently working as Professor in Department of Electronics and Communication Engineering. He guided 1 Ph.D, 32 PG and 30 UG projects and supervising 02 Ph.D students. He has published more than 50 papers in international journals and conferences. He is a member of professional bodies such as IEEE, ISTE and IETE. He is presently a technical program committee member of nine IEEE international conferences. His research interests are in the areas of wireless mobile communication, cellular networking, distributed cooperative communication, MIMO and signal processing applications, mm wave applications and bio medical signal processing applications.

Pardhasaradhi Pokkunuri was born in India, A.P, in 1978. He received UG, PG and Ph.D degrees from Acharya Nagarjuna University, A.P, India in 1998, 2000 and 2012 respectively. From 2000–2012 he worked as lecturer in Hindu College, Machilipatnam, from 2012 to 2014 worked as Associate Professor in Sri vasavi institute of engineering and technology, Nandamuru and 2014 until the present date, he has been working as a Professor in Electronics and Communication Engineering, Koneru Lakshmaiah Education Foundation, Vijayawada. He has published more than 100 papers in international journals and conferences. He is reviewer for several international journals including Elsevier and Taylor & Francis and served as reviewer and co-chair for one international conference. His research interests include antennas, liquid crystals and its applications. He is a life member of IACSIT, IAENG, UACEE and IIRJC. He is an editorial board member for a number of indexed journals.

About this Book

An ever-growing demand for higher data-rates has facilitated the growth of wireless cellular networks in the past decades. Cellular communication systems often suffer from interference, fading and multipath distortion caused by multiple users sharing a limited range of frequency bandwidth. These networks, however, are known to exhibit capacity and coverage problems to the end-user. Nevertheless, many aspects of cooperative communications are open problems. Cooperative communication scheme is an inherent network solution, based on relay nodes, and has emerged as a promising approach to increase spectral, power efficiency, network coverage, and to reduce outage probability. Transmitting independent copies of the signal generates *diversity* and can effectively combat the deleterious effects of fading. In particular, spatial diversity is generated by transmitting signals from different locations, thus allowing independently faded versions of the signal at the receiver. Space diversity techniques are particularly attractive as they can be readily combined with other forms of diversity, e.g., time and frequency diversity, and still offer dramatic performance gains when other forms of diversity are unavailable. In contrast to the more conventional forms of space diversity with physical arrays, this work builds upon the classical relay channel model and examines the problem of creating and exploiting space diversity using a collection of distributed antennas belonging to multiple terminals. Although the advantages of multiple-input-multiple-output (MIMO) systems are well known, it may be impractical to equip very small mobile equipments with multiple antennas. This is primarily due to the size and power limitations of these nodes. To overcome these issues, and embrace the benefits offered by the MIMO systems, the concept of cooperative relaying can be successfully implemented in cellular communication.

Due to the advances in cooperative communication, many efforts have been spent understanding and improving the benefits of deploying cooperative relays in wireless networks. However, most of these studies on cooperative relaying have been mainly looked at from the standpoint of small-scale wireless networks typically a network with a single source, destination and relay which may not be realistic, especially for multi-hop wireless networks. On the other hand, in the last few years studies show that employing multiple channels in wireless networks can mitigate the negative effects of interference and thus substantially enhance the performance of wireless networks.

Motivated by these ideas, it is thus worth investigating the benefits of integrating multiple channels into large-scale cooperative wireless networks. In this book, we address this problem and propose a model known as *cooperative network with multiple channels* (CoopMC) which deploys cooperative relays in large-scale networks and uses multiple channels to reduce the impact of interference in such networks. Specifically, we focus on deriving the capacity of CoopMC model, including network capacity analysis, optimal power allocation, resource allocation and relay assignment and reveal the important insights on when a network can benefit from cooperative communications and how multi-channel networking can further improve the network capacity.

In Chapter 3, two sophisticated routing algorithms based on MACC and distance parameters are implemented to maximize the benefits of CR in a multi-radio-multi-channel large-scale wireless network. The combined throughput and transmission time of the proposed algorithms are similar to the IACR algorithm. The best relay node with the best channel capability is identified among candidate relay nodes by using nodes near to LOS between source and destination and away from the source node as a guide. The dynamic CA with high channel functionality is formulated at the access point using the global chart. According to the simulation, the proposed Cooperative relay selection algorithm reduces propagation delay and average hop count by 27.5 and 32.5 percent respectively.

Chapter 4 proposes an energy-efficient transmission scheme for large-scale multi-radio multi-hop cell ad hoc networks to reduce energy usage. Better BER efficiency can be accomplished without increasing network power by introducing spatial multiplexing. The suggested EET scheme utilizes spatial multiplexing and dynamic power allocation to the nodes to achieve the target BER. The proposed scheme saves about 99 percent of energy as opposed to fixed power transmission, according to the simulation results. The efficiency of the proposed EET is often evaluated in terms of the number of hops for a normalized distance between the source and destination. According to the simulation, we will save up to 91 percent of energy while we use the EET.

A hybrid multi-hop cooperative routing algorithm for LC-MANET was described in this work. To counteract the mobility impact and reduce the total number of hops, combine clustering and location-based techniques. Incorporated optimization mechanisms in each hop and obtained the optimum number of cooperative nodes by maximizing the number of transmitters and receivers simultaneously. When opposed to the conventional routing approach, simulation results indicate that the proposed algorithm saves up to 53.42 percent in energy consumption.

1

INTRODUCTION

1.1 Background

Wireless networking is broadly accepted today due to the ability for uninterrupted communications and mobile communication, and these solutions have changed the existence of knowledge transmission in wireless networks and will have an effect on wireless networks in the future. Technological advances and state-of-the-art networking capabilities have recently enabled customers to monitor the whole business setup remotely from anywhere. Electronic gadgets develop intelligent functionality and capabilities for domestic use and are powered by hand-held devices that make homes smarter. The concept of integrating wireless networking in existing networks has advanced; the Smart Grid, for example, uses a wireless connection to link computers and/or appliances. Remote learners have been taught by online video conferencing, webinars, and video interviews. Furthermore, in a large range of applications, wireless sensors are found in mobile devices that are simple to set up.

Moreover, wireless communication has already evolved as an integral part of daily life. Since wireless communication is everywhere and well known, it has been accepted as a basic means of communication as it comes to 4G wireless access, WLAN communication available at the office, at home or at any public place based on Wi-Fi or portable M2M communication, for e.g. tracking systems in the manufacturing industries and transport, GPS in cars, or medical devices. With various environmental variables, wireless users face undesirable wireless channel impairments, including flickering, path loss, scattering, reflecting and shadowing. But apart from all these channel impairments, there's also the fading effect to consider, which all degrades the effectiveness of

DOI: 10.1201/9781032714011-1

1

wireless communication. However, by the turn of the century, the progress of wireless communication had been slowed by various effects, including multi-path fading, path distortion and shadowing. Because of the variations in channel quality in the wireless environment, traditional wired communication techniques are difficult to use in the wireless world. People have embraced diverse approaches to leverage diversity in various areas of spectrum, such as time, frequency and distance.

New proposals to fix the problems of interference and fading are suggested in the sense of the design specifications for wireless communication. In the past, solving these effects has led to few trade-offs in terms of efficiency, but today there's a lot of focus on taking advantage of these characteristics to produce the optimal output. The transmission nodes/devices of the network take the benefit of relaying on cellular networks. Theoretically, working on a way to upgrade broadband networks may enhance wireless connectivity.

Bit and power allocation policies should be formulated over frequency, time and space as a result of channel knowledge, so that more and more resources are allocated to channels that are more effective and energy-saving by avoiding transmission to weak channels in order to conserve resources. While data on channel information is not available, space-time and space-frequency codes can be used to improve reliability. For improved spatial diversity, multiple antennas must be integrated into a modern wireless transceiver due to advancements in the MIMO Principle. However, in some applications where the size and cost of wireless equipment are limited, multiple antennas cannot be mounted in a single terminal. When collaborating with neighbor nodes in the wireless network, this has become a promising and enticing option. This collaboration is the result of a process known as cooperative communication.

1.2 Need and Importance

In an attempt to address long-accepted limitations, communication across wireless networks using relay nodes is now one of the most important innovations in recent times. Relay nodes, which include both the source and the receiving point, are an incredibly essential part of the cooperative process to improve output. As shown in Figure 1.1,

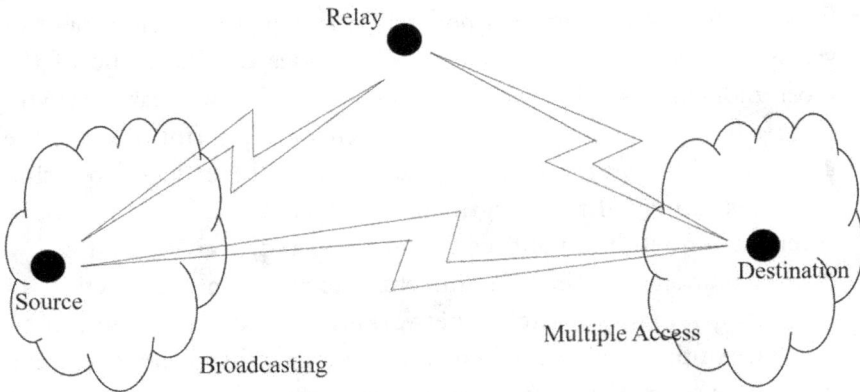

Figure 1.1 Simplest form of relay channel with three nodes.

the basic concepts of the relay network, especially the relay channel, have been applied to determine the transfer potential of the channel. The sum of bandwidth retained is the primary design factor in the relay network. Prior to the revival of recent interest in collaboration and relay networks, the research of relay networks was limited.

Cooperative networking, a type of network node sharing, improves the channel's congested capacity while also extending coverage and performance. The relay selection method, which makes use of the right estimate of the channel, also has the capacity to contribute to solving the above wireless channel problems. In the past decade, wireless communication has taken enormous strides. Although wireless system advances are dated to several years, major improvements are noted in data rates, hardware capacity, and battery life. New technological advances have made it possible for researchers to think outside the box and to build protocols for the wireless network that vary from conventional point-to-point (P2P) communication with a central control node. The ad hoc and cooperative wireless networks are an example, where the term "any" is omitted from the canonical taxonomy, enabling wireless communication for any node. The wireless channel's competitive architecture is what attracts new research interest and is the most significant part. This means that this nature of broadcasting is infringed where the propagation from the source node is unable to reach the expected destination because, for that reason, there is a fair benefit. In order to generate new ideas about distributed communication, the flow of information between nodes is essential. The advancement of various MIMO-enabled antenna technologies has been one of the most momentous scientific developments in recent years. MIMO devices have the greatest capability to increase the performance of the network using signal processing techniques. These data-processing strategies can entail the integration of signals obtained from various antennas as a result of multiple transmitting and receiving routes.

Cooperation is a fresh architecture requirement in reaction to the ongoing trend in wireless networking that uses the transmitted existence of the wireless medium and achieves transmitting through a distributed setup. While this new region promises advantages such as a reduction in power usage, an increased service area and enhanced data speed, additional problems need to be thoroughly explored.

1.3 Motivation

The higher demand for data rates in cellular networks has precipitated the proliferation of these wireless networks in the past decades. Cellular networking networks are vulnerable to interference, multipath distortion and fading due to shared frequency bands. However, these networks face bandwidth and coverage challenges for the end user. Regardless, most knowledge about cooperative cooperation remains an unanswered topic. Cooperative networking scheme is an underlying network solution, based on relay nodes and an approach to improve spectral, power performance and network coverage. The variety in the propagation of the signal will minimize the effects of fading.

Spatial diversity is created by spacing out the signals at the receiver, letting each receiver receive a slightly different version of the transmitted signal. Space diversity approaches are especially appealing because they provide a wide variety of diversity options without depending on the availability of other sources of diversity, e.g., time and frequency diversity. This cooperative network formulates the problem of producing and optimizing space diversity across a series of distributed antennas belonging to many terminals, as opposed to conventional forms of space diversity with specific physical arrays. While MIMOs are known for their various advantages, it may be impractical to equip small mobile equipment with several antennas.

This is because of the power needs of these machines. In order to solve these problems, and accept the advantages of MIMO cell systems, cooperative protocol relaying is the way to go.

In the advancement of cooperative communication, numerous attempts have been made to explore the use of cooperative relays and to improve the benefits of their use. However, most of these mutual relay experiments have been limited to local wireless networks and are typically a single source, destination and relay network that is inefficient to multi-hop wireless networks. Multiple channels within wireless networks may reduce a few of the negative consequences of interference, according to new research, enhancing network efficiency.

1.4 Problem Statement

Enabling co-operation in wireless networking ultimately relates to the nature of relay collection and cooperative routing concerns. While all of these architectural issues have been discussed separately and thoroughly over the past few years, the design of cooperative networks still has some problems that need to be solved.

- Extending the cooperative communication technique to large-scale networks is prone to significant interruption; multi-radio multi-channels are used in large-scale wireless networks to minimize interference by concurrent transmissions over orthogonal channels. Capitalizing on these possible advantages, there is a need for efficient routing and powerful mapping of radio channels.
- Energy efficiency is a crucial factor in the success of large-scale ad hoc wireless networks. In a cooperative network, by incorporating spatial diversity, the performance of the network can be improved with the same amount of power and energy consumption. There is also the need to minimize energy consumption with the optimal power allocation approach.
- Cooperative Communication (CC) is implemented in large-scale mobile ad hoc networks to leverage the benefits of the CC technique. Energy consumption and network lifetime are major challenges for large-scale Cooperative Mobile Ad hoc Networks (LC-MANET). The high data rate applications are

rapidly requiring high capacity, which increases the energy usage and thus reduces the network's lifetime. However, routing is also a key issue in MANETs, since every node in the network has mobility so it can freely move in any direction. Therefore, there is a need for an effective routing algorithm to improve reliability of the network and energy consumption.

1.5 Objectives

The key aim of this research is to examine the effects of incorporating multiple platforms into large-scale, cooperative wireless networks. The objectives of this work are set out below:

1. To design and analyze the relay selection criteria for faster data transmission in cooperative networking with multi channels.
2. To formulate an optimal solution for energy-efficient transmission for large-scale cooperative networks with multiple channels.
3. To propose the optimal routing mechanism for large-scale cooperative networks by considering mobility factor.

1.6 Methodology

The methodology given below is a step-by-step procedure to solve the considered application. From Figure 1.2, it can be noticed that firstly the specifications of cooperative network with multiple channels are to be specified based on the literature survey. Secondly, based on the specifications, the objectives are listed. Later, the proposed cooperative network with multiple channels is designed and simulated with the help of optimization techniques to meet the required specifications. Finally, the proposed cooperative network with multiple channels is implemented to improve its efficiency.

1.7 Aims and Significance

Cooperative networking is among the important dynamic strategies for solving modern wireless network limitations, and it is projected to

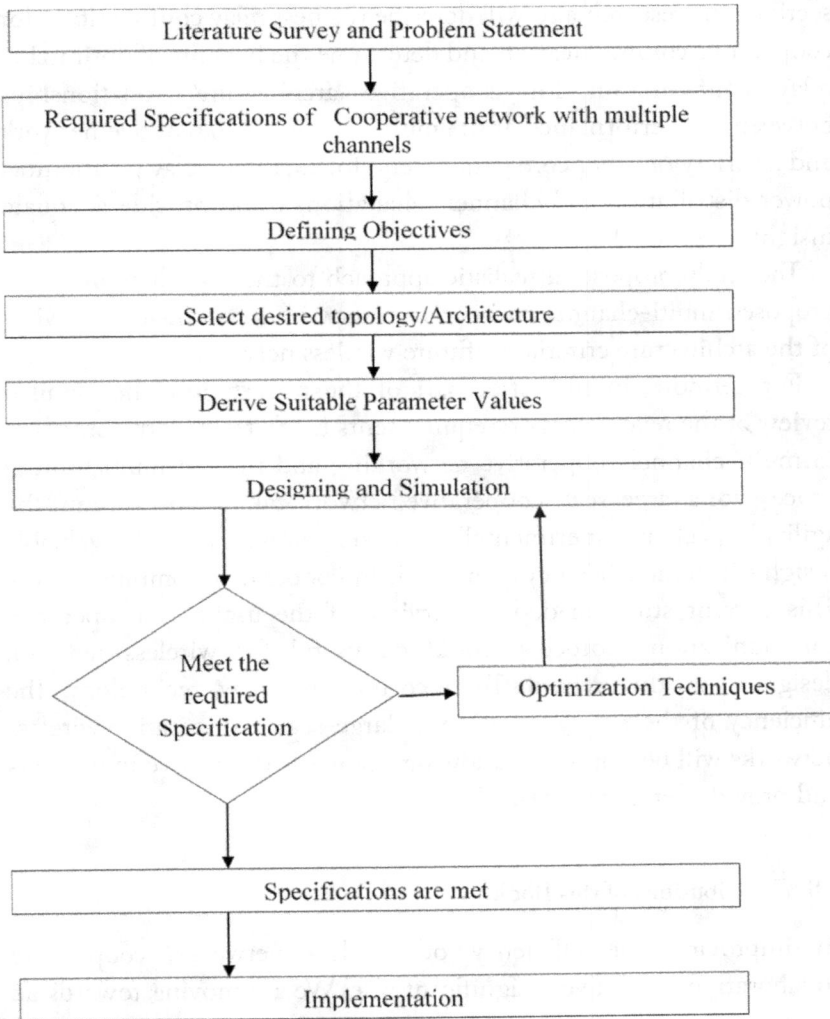

Figure 1.2 Methodology adopted.

play an important role throughout the structural design of upcoming wireless network generations. Researchers are often involved in designing prototypes that are accurate, successful and adaptable.

The main goal of this research is to provide a theoretical analysis as well as a relevant simulation evaluation in order to improve cooperative communication as an effective thing for future wireless networks, thus attempting to resolve the research problems discussed in the previous section. The research also will describe the best relay configuration for cooperative communication and determine the benefits of optimizing relay configuration for cooperative architecture. Relationships between the performance of the multi-channel collaborative network and a variety of other core parameters (for example, relay positioning, power distribution and channel calculation) could provide adequate insight.

The study proposes a realistic approach to the introduction of the proposed multi-channel co-operative network for evaluation in view of the architecture criteria for future wireless networks.

Furthermore, in the latter part of the dissertation, the detailed review of the relay selection requirements for faster data transmission in multi-channel cooperative networking and the optimum routing process for a large-scale cooperative network, taking into account the agility aspect, is experimentally checked, providing some valuable insights into the efficiency gains made by cooperative communication. This current study in-depth overview of the usage of cooperative communication protocols would be useful for wireless network designers in this regard. By incorporating new technology, the efficiency of the new generation of large-scale, cooperative wireless networks will be improved, allowing the network to serve more users and provide better services.

1.8 Contributions of this Book

In improving the efficiency of wireless networks, cooperative collaboration plays a very significant role. We are moving towards an in-depth analysis of the design in order to understand the complexities of cooperative communication. This is achieved by looking at two main collaboration characteristics, optimal relay selection requirements and optimal performance optimization. In this section, we address the main contributions made to fill the information void. In Chapters 3, 4 and 5, these contributions are explored in depth and can be highlighted as follows:

- *Performance investigation of an effective cooperative routing algorithm for large wireless networks focused on capability and location.* In this analysis, an algorithm is suggested yet at the same time considering: (1) collection of relays using parameters of distance and maximum-available-channel-capacity; and (2) distribution of channels using a dynamic global table. Our algorithm selects a relay node based on the shortest-path channel bandwidth available to reduce transmission delay without impacting aggregate throughput. The assessment and review of the proposed model was compared with the interference-aware-cooperative-routing algorithm and found that the proposed algorithm increased the delay reduction by 27.5 percent and the total hops reduction by 32.5 percent.

• *Formulation of an optimal solution for energy-efficient transmission for large-scale cooperative network with multiple channels.* In this chapter, we suggest two schemes: (1) cooperative spatial multiplexing: improving BER performance; and (2) energy-efficient transmission: minimizing network energy consumption. The simulation results show that the proposed scheme saves nearly 99 percent of energy relative to the fixed power transmission. Further, the output of the proposed EET is analyzed for the number of hops with a normalized distance between the source and the destination. It is found from the simulation that we can save up to 91 percent of energy using the EET.

• *Design of the optimal routing mechanism for large-scale cooperative network by considering mobility factor.* This chapter suggests a hybrid multi-hop shared routing algorithm. Large-scale network MANET is improved by incorporating clustering and location-based routing techniques to improve network life. When the request flow arrives, the network is split into clusters by the cluster header. The clusters under the base station form a location-based community of several cluster members. After the community is created, one of the nodes is chosen as the cluster header for the location involved, which will improve the existence of the network. Energy-efficient hybrid cooperative routing algorithm for LC-MANET is used to maximize the number of transmitters and receivers in each hop. Then achieve an optimum number in the cooperative nodes to minimize the end-to-end energy consumption. The assessment result indicates that the proposed algorithm saves up to 53.42 percent of energy consumption relative to conventional algorithms.

2

COOPERATIVE COMMUNICATION

2.1 Background

Wireless networking is broadly accepted today due to the ability for uninterrupted communications and mobile communication, and these solutions have changed the existence of knowledge transmission in wireless networks and would have an effect on wireless networks in the future. Technological advances and state-of-the-art networking capabilities have recently enabled customers to monitor the whole business setup remotely from anywhere. Electronic gadgets develop intelligent functionality and capabilities for domestic use and are powered by hand-held devices that make homes smarter. The concept of integrating wireless networking in existing networks has been extended, such as the Smart Grid, where a wireless connection will link computers and/or appliances. Remote learners have now been taught by online video conferencing, webinars and video interviews. Furthermore, in a large range of applications, wireless sensors are found in mobile devices that are simple to set up.

Moreover, wireless communication has already evolved as an integral part of daily life. Since wireless communication is everywhere and well known, it has been accepted as a basic means of communication as it comes to 4G wireless access, WLAN communication available at office, at home or at any public place based on Wi-Fi or portable M2M communication, for e.g. tracking systems in the manufacturing industries and transport, GPS in cars, or medical devices. With various environmental variables, wireless users face undesirable wireless channel impairments, including flickering, path loss, scattering, reflecting and

DOI: 10.1201/9781032714011-2

shadowing. But apart from all these channel impairments, there's also the fading effect to consider, which all degrades the effectiveness of wireless communication. However, by the turn of the century, the progress of wireless communication had been slowed by various effects, including multi-path fading, path distortion and shadowing. Because of the variations in channel quality in the wireless environment, traditional wired communication techniques are difficult to use in the wireless world. People have embraced diverse approaches to leverage diversity in various areas of spectrum, such as time, frequency and distance.

New proposals to fix the problems of interference and fading are suggested in the sense of the design specifications for wireless communication. In the past, solving these effects has led to few trade-offs in terms of efficiency, but today there's a lot of focus on taking advantage of these characteristics to produce the optimal output. The transmission nodes/devices of the network have the benefit of relaying on cellular networks. Theoretically, working on a way to upgrade broadband networks may enhance wireless connectivity.

Bit and power allocation policies should be formulated over frequency, time and space as a result of channel knowledge, so that more and more resources are allocated to channels that are more effective and energy-saving by avoiding transmission to weak channels in order to conserve resources. While data on channel information is not available, space-time and space-frequency codes can be used to improve reliability. For improved spatial diversity, multiple antennas must be integrated into a modern wireless transceiver due to advancements in the MIMO Principle. However, in some applications where the size and cost of wireless equipment are limited, multiple antennas cannot be mounted in a single terminal. When collaborating with neighbor nodes in the wireless network, this has become a promising and enticing option. This collaboration is the result of a process known as cooperative communication.

2.2 Wireless Communication

Guglielmo Marconi invented wireless communications in 1895 when he used electromagnetic waves to send a three-dot Morse code for the letter "S" across a distance of 3 kilometers. Since then, wireless communication has evolved dramatically in terms of integrated circuitry and technical developments, culminating in today's extremely sophisticated networks. The never-ending drive for better throughput, higher dependability, higher data transfer and cost-effectiveness has inspired everything from satellite transmission to radio and television broadcasting to the current development of 4G for mobile communications. Exploiting technical advancements in radio hardware and integrated circuits, which allow for the implementation of increasingly complex communication schemes, would necessitate an assessment of wireless networks' fundamental performance constraints.

A wireless network system may be thought of as a collection of nodes attempting to interact with one another. However, because wireless channels are broadcast, those nodes might be thought of as a network of antennas scattered across the wireless system. The signal transmission between these antennas suffers from significant deterioration, prompting extensive study into how to successfully overcome these detrimental consequences. For a better understanding of the cooperative communication technique, some of these channel difficulties will be detailed.

Multipath fading causes mistakes and distortions, and compensation methods may be divided into three categories: forward error correction, adaptive equalization and diversity algorithms. To fight the error rates experienced in a mobile wireless environment, approaches from all three areas are often employed.

2.3 Glance of Wireless Channel

We'll go through the various styles of wireless communication and their effectiveness and impairments with their specific applications and results in this section.

2.3.1 Phenomena of Wireless Propagation

Multi-path propagation of wireless communications can lead to improved results if properly utilized. This effect occurs as the radio waves follow different tracks from the starting point to the ending point. As the signal is transmitted on the radio channel it is reflected, refracted, diffracted, dispersed and absorbed also by multiple (delayed

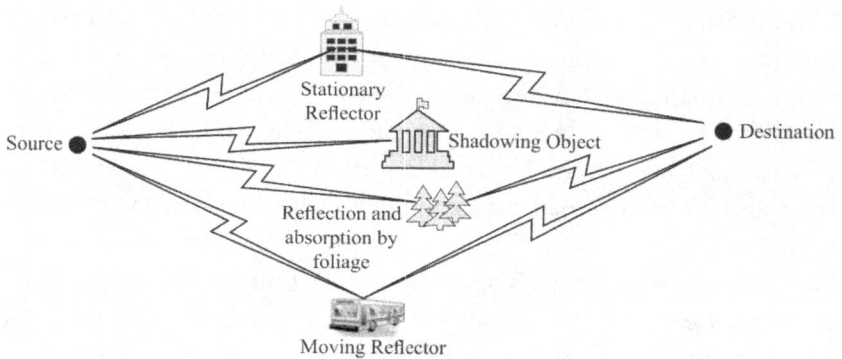

Figure 2.1 The propagation and impairments of Wireless Signal.

and attenuated) radio waves at the end point. The wireless signal spread and its impairments are displayed in Figure 2.1.

A further effect is produced when the wave-front of the incident returns to the medium it came from. A reflection is a basic effect that happens when a propagating electromagnetic wave interacts with an object which is large enough in comparison to the propagating wave frequency. Reflection can be classified into three types.

1. Specular smooth surface reflection;
2. Reflections from rugged texture; and
3. Physical optic reflections.

Both moving and stationary objects can cause reflection. The partial blockage by surfaces with uneven edges of the electromagnetic wave front in relation to reflection results in diffraction. As a result, a wave bending around the barrier is observed, even though the LOS path between the target and the source is not clearly visible. Models are used to describe diffraction in two ways: the knife-edge and wedge-diffractions.

In the end, unusual dispersion occurs where the electromagnetic wave wavelength exceeds the specifications of the devices which interrupt the radio channel. Radio waves are greatly influenced by the rough surface of the Earth, thus scattering the energy in all directions. Electric car scatters, trees, fog, signs, and road signs are examples of wireless communication scatters. The trees, the fog, the signboards placed on the streets and reflectors lighting up the road in the evening time are some examples of wireless communication scatters.

The essences of these results appear in varying patterns depending on the situation of the channel and objects in the way that block the signal to its target. These wireless phenomena have both some gains and impairments.

2.3.2 Impairments in Wireless Channel

The basic wireless channel and the associated degrading process should be known to understand wireless communication. Specifically three major impairing attributes, i.e., path loss, shadowing and fading, are to be taken into consideration for the signal propagation over a wireless channel.

At a certain distance from the transmitter the average power obtained from the receiver results in a loss of power or Pathloss over that distance. Most of this effect relies on linearity as it can reduce unwanted noise interference well. Shadowing leads to deadlock in contemporary communication networks because shadowing results in connection outages. This result in the varying power level is achieved at any distance from the transmitter. Shadowing in both nature and human nature can be described as mysterious and spontaneous and is historically modelled in decibels as Gaussian.

Finally, the attenuation of a particular wireless channel affecting a signal is fading. With time, frequency and geographical location it can vary. The true essence of the wireless channel as fading is discovered by allowing the averages Pathloss and Shadowing signals to waver. Due to multipath propagation, this is really a constructive/destructive expansion of the measured signal. Because of the instantaneous nature of this effect, this is usually referred to as fast-fading and slow-fading. Present wireless communications networks may either be rapid or slow to fade, depending on the level of mobility. When signal copies (symbols or bits) are exposed to multipath propagation, they are coupled in a frequency-selective fading channel to detect a signal if the symbol length is larger than that of the reciprocal propagation time delay If the propagation time is less than the symbol range, flat or merely flat wavelengths, the signal is observed.

Flat fading and frequency-selective fading channels could also be defined depending on their coherence bandwidth. The reciprocal delay propagation is characterized as the duration of the delayed signal, duplicated by several pathways to the destination, quantifying the rate at which non-correlated fading can occur. The signal bandwidth of the flat fading channel is greater than that of the consistent channel, and conversely for both the frequency selective channel. Multipath fading is used to highlight variations in envelope voltage using one of the widely used probability distribution functions (pdf). The amplitude of the envelope distorted by multi-way components as well as a LOS part can be used for a probability density function. We'll go through the most critical fading processes.

- *Rician Fading:* Whenever a non-fading signal section like the propagation of the LOS path is present, a Rician system is considered. As a result, wireless channels with little dispersion as well as flat terrain will be found in Rician fading features. The Rician distribution's PDF signal is given in the model, which includes a direct path and a dispersed wave:

$$f_h(h|l\sigma) = \frac{h}{\sigma^2} \exp\left(\frac{-(h^2 + l^2)}{2\sigma^2}\right) I_0\left(\frac{hl}{\sigma}\right) \qquad (2.1)$$

Where, h(h ≥ 0) is the fading channel amplitude, σ is called as the fading channel variance, l(l ≥ 0) is nothing but the LOS path and I0 is known as the Bessel function with order zero of the first kind.

- *Rayleigh Fading:* When the receiver has a relatively large number of components, the distribution is assumed (with equal power and separate phases). In addition, multipath fading is widely used without a clear LOS path. In this case, the distribution pdf of Ricians refers to the distribution of Rayleigh and is provided with the [1],

$$f_h(h\sigma) = \frac{h}{\sigma^2} \exp\left(\frac{-(h^2)}{2\sigma^2}\right) \qquad (2.2)$$

- *Nakagami Fading:* The distributions of Nakagami are considered more stable and tractable mathematically than the distributions of Rician because of the lack of a modified Bessel function. Nakagami distribution is a statistical concept without any physical foundation. However, unlike Rician and Rayleigh distributions, such findings are accurate because they are based on observations of physical quantities.

Signal strength might be represented in decibels (dB) at the receiver if the previously mentioned algorithmic notation impairments were represented as Pathloss (X), Shadowing (Y), and Fading (Z).

$$P_r = P_t + X + Y + Z \qquad (2.3)$$

These limiting variables add up to decibels, while they are multiplicative on a linear scale. To address these flaws, cooperative communication intends to deceive rather than mitigate the multi-way propagation effect, thereby maximizing the effect's benefits.

2.3.3 Forward Error Correction (FEC)

This is also known as forward error correcting code, and it occurs when the sender inserts carefully selected redundant material to its messages. The receiver may then discover and repair mistakes without having to ask the sender for more information. There are two sorts of FEC codes:

1. Block Codes
2. Convolutional Codes

- *Block Codes:* Block codes function with specified fixed-size blocks. It converts a message m, which consists of a sequence of information symbols throughout an alphabet, into a code word, which is a set length sequence s of e encoding symbols.
- *Convolutional Codes:* This is applicable to symbol streams of any length. A sequence of information bits is sent through a shift register, with two output bits created and communicated for each information bit. For each pair of two channel symbols it receives, the decoder calculates the status of the encoder. It can decode the original information sequence by knowing the encoder's state sequence.

2.3.4 Adaptive Equalization

A technique for preventing inter-symbol interference in transmissions of analogue or digital data. It entails a means of reassembling the scattered symbol energy into its original time sequence.

A linear equalizer circuit is a popular technique to adaptive equalization in which the input samples are individually weighted by coefficients that are dynamically modified depending on a training sequence of bits. The training program is broadcast. The receiver compares the received training sequence to the predicted training sequence and generates appropriate coefficient values based on the comparison. A fresh training sequence is delivered on a regular basis to accommodate for changes in the transmission environment.

It may be essential to incorporate a fresh training sequence with each block of data for Rayleigh fading channels. Again, this is a significant expense, but the mistake rates experienced in a mobile wireless context justify it.

2.3.5 Diversity

To increase dependability, two or more communication channels with distinct characteristics are used. Individual channels suffer independent fading occurrences, resulting in diversity. Various logical channels between the transmitter and receiver can so compensate for error effects, as can receiving multiple copies of the same signal, which are then integrated at the receiver. Because the transmission has been stretched out to prevent being subjected to the greatest possible error rate, this strategy does not eliminate mistakes, but it does lessen them. The many diversity strategies will be discussed since they provide the foundation for cooperative communication, which is the topic of this thesis.

2.4 Diversity Techniques

Several classes of diversity schemes have been identified which include the following:

2.4.1 Space Diversity

The signal is sent via numerous separate propagation channels in this strategy. Numerous receiving antennas (receive diversity) and/or multiple transmitter antennas can be used to achieve this (transmit diversity). Because each antenna will face a distinct interference environment, multiple antennas provide a receiver with several views of the same signal. As a result, if one antenna has had a profound fade, it is likely that the other will get an adequate signal.

2.4.2 Frequency Diversity

Transmission is carried out via numerous frequency channels or over a large spectrum that is impacted by frequency-selective fading in this situation. Because the wavelengths for distinct frequencies result in separate and uncorrelated fading characteristics, it includes the simultaneous use of numerous frequencies to transfer information. OFDM and spread spectrum are two examples of this.

2.4.3 Time Diversity

Multiple copies of the same signal are transmitted at various times to produce temporal diversity. Alternatively, before transmission, a redundant forward error correction code is inserted, and the message is structured in a non-contiguous format. Error bursts are thereby avoided, and error rectification is made easier.

2.4.4 Polarization Diversity

To transmit several copies of the signal, various antennas with different polarizations are required. At the receiver, a diversity combining technique is used to combine several received signals into a single better signal.

2.5 Diversity in Wireless Channels

The wireless channel now briefly displays diversity as a result of the diversity gain described in the earlier sector. Fading wireless channels may cause an increase in error rates and a fading slump on a single source and destination over time. This will necessitate implementing methods to reduce error rates, such as enabling automated error correction capabilities, lowering transmission rates, or using more sensitive detectors. This and other methods of prevention, on the other hand, may not be feasible. Taking into account the effects of wireless channel fade and multipath propagation, one can improve the efficiency of the connection by looking at multiple directions of reference and each displaying an individual fade as much as possible. When using this technique, the chances of getting a skewed signal are reduced, or producing multiple signal copies would gradually lower the rate of error.

These methods are known as diversity approaches, and they are divided into two groups: time and space. Numerous copies of the signal will be transmitted over various time periods, and multiple copies will be sent to various carrier frequency channels. Furthermore, as shown in Figure 2.2, spatial diversity employs several transmitter and/or receiver antennas, with diversity achieved in the spatial diversity (or antenna diversity).

When there are multiple antennas on the receiver, the SIMO systems and MISO method for installing more than one transmitter antenna are shown in Figure 2.3. Multiple channels are formed between antenna pairs as a result, as well as the receiver's ability to retrieve the signal. Given that the channels introduced are isolated or strongly un-correlated, the risk of connection loss often decreases.

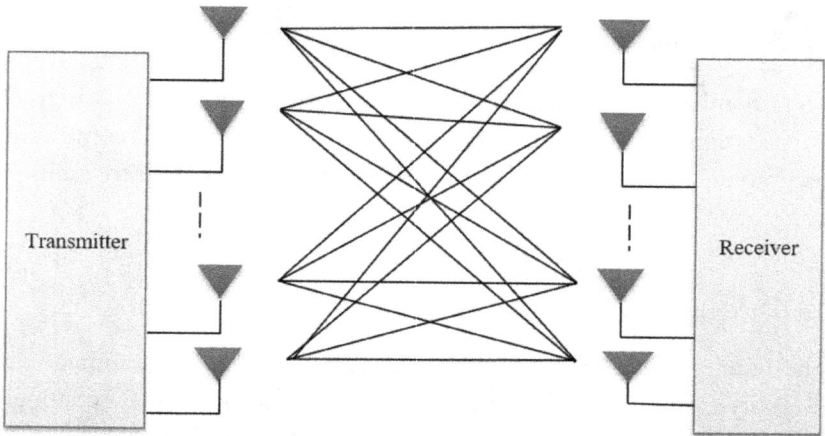

Figure 2.2 MIMO system.

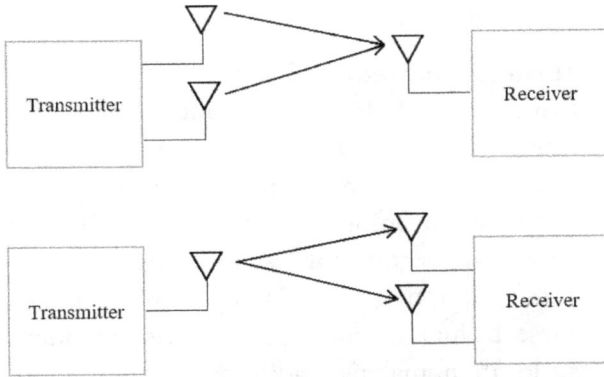

Figure 2.3 Transmit and Receive Diversity.

Increasing the different number of antennas increases the redundancy (diversity) of the signals received, which improves receiver detection efficiency, and the difficulty of signal recovery at the destination is also reduced with the number of antennas. As a result, constructive integration of signals arriving at the receiver from multiple paths is critical. As opposed to each actual signal, the aim is to integrate the different signals in just such a manner in which the resultant signal is of better quality or has less errors. Signal combining is dependent on certain design priorities and often relies on SNR. It is recognized as receiver diversity. Equal Gain Combining (EGC) is a procedure that is used in which all of the signal copies are done in the same way during the combining phase. In Maximum Ratio Combining (MRC), different signals are combined depending on weights, and Selection Combining (SC) has been used to pick the best version of the signal. After a MISO device has passed through individual fades, the received signal is a mixture of signals from all transmitting antennas. Transmitting diversity, which is based on a variety of transmission antennas, is a term used to describe such redundancy. Redundancy versions of the signal are commonly implemented at the receiver to increase transmission efficiency. However, using the Space-Time Bloc Coded (STBC) strategy is a rare strategy, which provides for diversity gains and reaches a maximum rate. This method typically provides a diversity advantage at a lower coding cost.

2.6 Cooperative Communication

Wireless communications is now a widely sought-after communication technology that is particularly useful for mobile access. It has gone through several development phases since its beginnings in order to fulfil the ever-increasing demands of its diverse range of applications. The impacts of multipath fading, shadowing, and path loss on wireless channels are the most significant issues in the history of wireless communications, prompting much study into potential remedies. Because of these factors, channel quality varies randomly in time, frequency, and location, making traditional wireline communication techniques challenging to utilize in a wireless context. Despite repeated proposals, high-efficiency approaches were never achieved until the introduction of diversification techniques in the last two decades.

Variety technology significantly enhances the performance of wireless communications by providing independent fading paths for signals during transmission, allowing them to utilize diversity in multiple channel dimensions such as time, frequency, and space, and so obtain diversity advantages. Advances in the theory of multiple-input multiple-output (MIMO) systems, in particular, have made it advantageous to equip contemporary wireless transceivers with numerous antennas for spatial diversity advantages. However, having numerous antennas on a single terminal is impracticable for many applications due to the size and cost limitations of wireless devices, such as wireless sensor networks or cellular phones. In such instances, the simplest and most promising solution is to create a virtual MIMO environment in which nodes collaborate and share their antennas to build a distributed virtual MIMO antenna system. Cooperative communications are used to accomplish this.

Furthermore, the introduction of 4G mobile communication has resulted in heterogeneous networks of diverse services that employ multiple standards and, as a result, different terminals to deliver services. The strategy of deploying network services using a single all-purpose device causes design issues, resulting in wasteful battery power usage and limited battery life. In these scenarios, cooperative communications allows users to reduce network load and so extend the capacity and battery life of their devices.

We're discussing architectural criteria and network dynamics to help us discuss about flaws and make wireless channels more relevant. In cooperative systems, the technical breakdown of many methods is taken into account.

2.6.1 Transparent and/or Regenerative Relaying

A design which needs transparent or regenerating relaying is typical for cooperative systems. Prior to retransmission, transparent relaying amplifies the signal. A digital domain relay, on the other side, will process and retransmit a signal using regenerative relaying. In the forthcoming discussion, a thorough illustration of protocols adopted for open and regenerative relaying will be explained.

2.6.2 Dual-Hop and/or Multi-Hop Networks

The total of relay stages is another significant parameter to remember. Obviously, relays may be built in sequence or in parallel; we must prioritize having multiple relay stages over additional complexity and a variety of other parameters. The advantage of Pathloss is that it allows you to increase the number of hops you take (in series). But at the other side, increasing the number of relays increases the overall order of diversity (in parallel). Any architecture refers to use of interference reduction methods or orthogonal or temporally isolated relay transmissions because of the added complexity.

2.6.3 Direct Source to Destination Link

Since wireless communication often involves the loss of a direct connection from the source to the destination, transmission can depend on a relay signal (s). A limited capacity wireless device is only accessible if a direct source to destination connection is available, whereas limited range systems are not.

The wireless communication channels as well as the mode of operation influence the architecture. Implementation of these principles into a more functional cooperative structure to address the demands of real-world wireless communication networks is also in the process.

2.6.4 Advantages and Disadvantages of Cooperation

We will now list several advantages and disadvantages of using cooperative architectures, which were briefly listed in the previous discussion. The inherent performance gain is one of the benefits of utilizing cooperative designs. Optimal resource control and coordination can help to increase overall system efficiency. These gains can be interpreted as increased bandwidth, reduced transmitting forces, or increased cell coverage, all of which contribute to improved service quality (QoS). Cooperative designs, on the other hand, take advantage of existing infrastructure with minimal modifications in order to increase efficiency. In a disaster-affected region with limited coverage, cooperative connectivity using such low-cost designs without networks can be successful. In the meantime, it has been stated that emerging cellular networks will coalesce with cooperating relay nodes, leading to lower operation and maintenance costs. Cooperative designs, on the other hand, may include a variety of contingencies.

The best relay for the cooperative architecture must meet some requirements relevant to the diverse wireless channel, which adds to the design process's difficulty. In cooperative systems, this difficulty is usually higher than in non-cooperative systems. Increasing the number of relays in the cooperative stage, on the other hand, could introduce difficulty, overhead, or interference, eroding the physical layer performance advantage obtained from avoiding higher-level measures like the MAC layer or the Network layer. As a consequence, more sophisticated and practical scheduling algorithms are needed. A close synchronization device can also be regarded to facilitate collaboration since the transmission mechanism depends on the addition of cooperative relay nodes. As a result, signal synchronization methods and associated hardware are required to achieve synchronization. Finally, for reliable and precise distribution of channel estimates, a field of research with the numbers of relays and related WLANs is required, which expects significant input from researchers and thus provides an incentive to develop cooperative designs to smoothly transition away from current non-cooperative designs.

Although the disadvantages of cooperative designs outweigh the benefits, careful system design and preparation are critical for ensuring optimum efficiency and, as a result, limited system performance degradation when using cooperative transmission systems.

2.6.5 Cooperative Performance Bounds

The realistic limitations of effectiveness in this sub-section will be discussed briefly, taking into account the benefits and drawbacks of cooperative architectures. As a result, the highlighted hardware architecture and similar protocols based on the OSI model are affected by these efficiency metrics. We will demonstrate the different wireless channels and their effects on possible rate benefits, and DMT in the context of co-operative systems in this section by applying those efficiency restrictions without going into great detail.

2.6.5.1 Capacity Gain in Ergodic Channel   Ergodic channels are really a form of process that requires averaging Shannon power over time. In terms of channel variety, Ergodic channels enable transmission to traverse all fading states. As a result of ergodicity, the averages concept can accomplish equivalent and simultaneous average information over infinitely long code words. An Ergodic channel could even maintain confidence while transmitting at the maximum possible rate.

$$C = \mathrm{E}_h \left\{ \log(1 + \gamma) \right\} \tag{2.4}$$

Where h denotes the gain in channel/power which is related to the instantaneous signal to noise ratio (SNR) as $\gamma = h\dfrac{S}{N}$. $\mathrm{E}_h\{\bullet\}$ denotes the expectation operator. Considering, for example, the flat Rayleigh fading with a PDF,

$$p_h(h) = \frac{1}{\bar{h}} \exp^{\frac{-h}{\bar{h}}} \tag{2.5}$$

and the average capacity can be evaluated as,

$$C = E_h \left\{ \log(1+\gamma) \right\}$$

$$= \int_0^\infty \log(1+\gamma) \frac{1}{h} \exp^{\frac{-h}{h}} \, dh \tag{2.6}$$

2.6.5.2 Rate Outage Gain in Non-Ergodic Channel Shannon's channel-capacity theory ignores conditions where communication criteria are different. It is possible to create a non-ergodic channel in which the transmission duration must fade almost continuously. Videlicet is a non-ergodic method in which the signal is precisely detected despite the time difference between samples. In practice, a slow fading channel shows this effect. As a result, the ultimate transmission rate R is not achieved through a non-ergodic channel, but rather through the management of a given rate R with a certain outage probability rate $P_{out}(R)$.

$$P_{out} = P\left(\gamma < \left(2^R - 1\right)\right) = \int_0^{2^R - 1} p_\gamma(\gamma) d\gamma \tag{2.7}$$

Where *SNR's pdf* is given by $p_\gamma(\gamma)$, $P(\bullet)$ is probability. For Rayleigh fading channels, the outage probability is,

$$P_{out} = 1 - \exp\left(\frac{-\left(2^R - 1\right)}{\bar{\gamma}}\right) \tag{2.8}$$

From Equation, the probability of outages decreases with increasing SNR can be readily observed.

2.6.5.3 Diversity-Multiplexing Trade-off (DMT) As previously noted, taking the integration scheme into consideration increases diversity. Random variations in signal level in time, distance, and frequency occur with wireless connections, which have a substantial influence on signal quality. Diversity means that multiple transmission versions are obtained at the destination from the source. In case of an isolated fading, the likelihood of the fading will be reduced when copies are mixed in the receiver because of variable channel conditions. This diversity increases the wireless system's efficiency. On the other hand, to maximize data efficiency, multiplexing or multiplexing gains are achieved via multiple independent signals transmitted simultaneously across spatial channels. This results in an enhancement in capacity without the need for more bandwidth or power. To enhance the effectiveness of a cooperative communication model, it is essential to include both types of gains in the design. However, it is not possible to achieve both simultaneously, and there is a basic tradeoff about the extent to which the advantages are sufficiently promising in the specific model. Since its debut, DMT has been acknowledged as a benchmark prospective choice. As previously stated, it enables for a basic compromise between diversity gain and multiplexing rate, or, to put it another way, the likelihood of disruption is reduced when a rate of contact and SNR are used. DMT can be used as a measure of efficiency for a wide variety of communication networks operating on a varied channel. Taking into account the word DMT, the diversity gain can be formulated again as,

$$g_d = -\lim_{\gamma \to \infty} \left(\frac{\log P_{out}(R, \gamma)}{\log \gamma} \right) \qquad (2.9)$$

Where, $P_{out}(R, \gamma)$ is the Shannon Power Expression for the probability of outage for a given average SNR, g_d is the gain in diversity. At high value of SNR multiplexing gain can be expressed asymptotically as [2].

$$g_s = \lim_{\gamma \to \infty} \left(\frac{R(\gamma)}{\log \gamma} \right) \qquad (2.10)$$

2.7 Cooperative Communication Design Aspects

The functionalities linked to cooperative networks such as relay nodes depending on the strategy/protocol for inherent transmission and the canonical flow of knowledge by using relay and associated

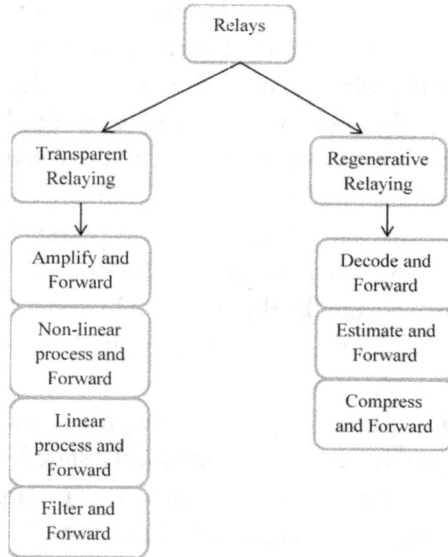

Figure 2.4 Relaying Taxonomy.

architectural metrics can now be exemplified. Figure 2.4 initially shows the taxonomy of the transmission node.

2.7.1 Relay Family/Type

In literature, there are several common methods to interact. Cooperating relay nodes are loosely split into two classes based on signal processing at relay nodes: transparent relaying and regenerative relaying.

- *Transparent transmission:* relay nodes typically do not change the received signal waveform information and are intended for basic analog signals, such as amplification or phase rotation. The analog signal is transmitted automatically via amplifiers and frequency conversions, as it is not encrypted in a digital domain. One of the most common so-called transparent relayed strategies/protocols is:
  - *Amplify and Forward (AF):* This is the most common method for amplifying the signal received by a relay prior to transmission. The amplification factor may be modified

using this technique so that the signal is better detected in the destination [3].

- *Linear and Forward Process (LF):* This method requires linear phase shifts to extend the analog signal. Due to the linear shifting phase [4], this technique can be used to create scattered beams.
- *Nonlinear and Forward (NLF):* This method is often called nonlinear AF [5], requiring non-linear treatment of the obtained analog signal. As a result, the end-to-end error rate induced by noise propagation is normally reduced.
- *Filter and Forward (FF):* This method for basic relay is intended for orthogonal multiplexing frequency-division transmission (OFDM). The FF strategy filters a analog signal directly and transmits the filtered signal to the location to minimize the processing difficulty inherent in the OFDM architecture [6].

The option of amplification factor and output strength is a critical design metric of transparent relay protocols. The node includes a constant amplification factor that is calculated by the channel's measurement over a specified time span in the case of fixed gain amplification. For example, based on the average canal gain between source and relay, the amplification factor is largely inverse. In non-ideal channel size, the amplification factor is usually high, leading to higher power output allocations. Since the maximum output power limits the retransmission, cutting effects will arise, necessitating mitigating techniques if the amplification factor is calculated. The amplification of the instantaneous channel gain value is approximately inversely proportional to the gain value of the channel between both the source and the relay throughout the case of variable gain amplification. Although the objective of a transparent relay is to maintain constant power, which is normally a hardware-related function, the requirements for amplification result in a suboptimal position to keep the power constant.

- *Regenerative relaying:* The information in the signals received by relay nodes in this family is altered, and more efficient hardware is required to perform complex operations. For this reason, regenerative relays typically outperform translucent relays. One of the most important regenerative relaying strategies/protocols is:

- *Decode and Forward (DF):* DF decodes a signal and then codes it before it is transmitted, in contrast to the Translucent AF technique. DF is recorded over a wide variety of application scenarios to exceed other loss and probability of error techniques [7].
- *Estimating and Forward (EF):* The relays amplify the signal using a signal detection algorithm, then collect the signal in a specified order of modulation and translate it to baseband using that technique. The measured signal is then retransmitted with the same or different module order as before [8].
- *Compress and Forward (CF):* This regenerative method can be re-encoded for retransfer, in some way connected with the EF strategy. This entails pre-transmission signal sample source coding which has been found to function well while the relay node is closest to the destination or when the relay to destination channel conditions are favorable [9].

Now that the relaying characteristics have been identified, the communication through the cooperative system should be explored. To this end, we will address the various communication flows that arise from the adoption of cooperative relay systems. In addition, we will address essential architecture parameters which contribute to the success of such networks.

2.7.2 Canonical Information Flows

Information usually reaches the destination via a direct connection in wireless communication networks. However, we can build three separate wireless transmission settings, given the cooperative architectures. In Figure 2.5, these configurations are shown,

- *Direct link:* This is really a basic wireless communication, as seen in Figure 2.5(a) that does not involve any cooperative relay nodes and enables the source to connect directly to the destination.
- *Serial relaying:* Serial relay facilitates the transmission of data across a relay chain (that can use orthogonal channels for relay information) between the source and destination as seen in Figure 2.5(b). Please note that a direct source to the

(a) Direct Link

(b) Serial Relaying

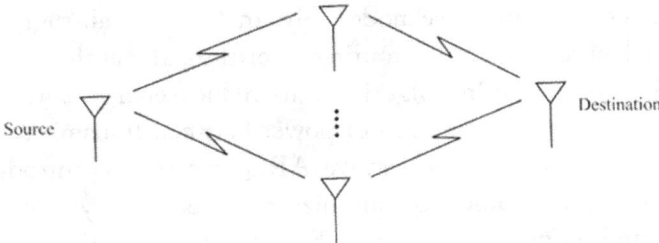

(c) Parallel Relaying

Figure 2.5 Canonical information flow by using (a) Direct Communication, (b) Serial relaying, and (c) Parallel Relaying.

destination connection may or may not be available in these and subsequent situations.

- *Parallel relaying:* Figure 2.5(c) demonstrates parallel relaying, which provides the transmission of parallel relays between source and destination. Network setups may be infrastructure-based or infrastructure-free. They may be operated by a central node or without one in either situation.

2.8 Background on Cooperative Routing

The Cooperative Communication (CC) Strategy of the day is now being introduced in nearly all wireless networks in order to achieve the benefits of Multiple Input Multiple Output (MIMO). Since the positioning of the relay affects the efficiency of the wireless networks, most research has been performed in the selection of relays, and it is found that the mid-located relay delivers better performance.

CC can be further applied to Large Scale Networks to boost network capacity. Network energy reliability is a crucial consideration in the choosing of cooperative relay nodes (Routing) to enhance the existence of mobile nodes/networks. Auon Muhammad et al. have introduced a power-conscious cooperative routing algorithm by implementing a super node concept. The authors Jesus Gomez-Vilardeb suggested an efficient minimum energy routing by giving the power rate ratio as the weights to the path. In order to achieve the lowest energy consumption, Xiangling Li suggested a routing algorithm based on the number of transmitting and receiving antennas and the set of cooperative nodes. Siyuan Chen et al. suggested an energy balanced cooperative routing algorithm. The authors proposed a MIMO-based routing algorithm to reduce energy consumption through a rational distribution of power between transmitters. Auon Muhammad Akhtar, a cooperative ARQ system was introduced at the MAC layer in order to minimize the total energy usage of the network and proposed an energy-efficient routing algorithm based on ARQ. Charles Pandana et al. (2006) suggested cooperative routing algorithms focused on the maximum power allocation. L. Zheng et al. (2011) suggested energy-efficient cooperative routing with truncated ARQ with an end-to-end throughput constraint to minimize the cumulative energy consumption.

Energy usage is a crucial problem for MANET. Owing to high data rate applications, there is an increasingly growing need for high bandwidth, which can further boost energy consumption, shorten network life and increase reliability. Cluster routing strategies overcome these challenges. One of the most common clustering systems is the hierarchy of lower adaptive energy clusters (LEACH). A variety of improved LEACH routing schemes have been proposed in recent years with a focus on modified cluster heads (CHs), network topology and network extension. In [10], the authors suggested a simulated cooperative MIMO transmission system and obtained an analytical expression for an optimum number of cooperative nodes for two-stage cooperative networks. A low-complexity cooperative routing algorithm was suggested in [11] and an optimal power allocation approach was introduced. To reduce network resources, the authors of [12] proposed routing algorithms by optimizing the efficiency of Physical, Medium Access Control and Network Layers. To this end, the authors suggested a cooperative automatic repeat request (ARQ) process on the MAC layer. A cooperative routing algorithm based on quality of service was implemented in [13,14] to minimize energy consumption. However, both of the above-listed authors identified that all the nodes are fitted with a single radio terminal.

The authors of [15] suggested an Opportunistic Cooperative Packet Transmission (OCPT) system for multi-hop cooperative networks. In OCPT, the cluster head chooses the transmitter and the receivers to form the MIMO before the transmission. Due to multiple transmitters and receivers in each hop, the energy usage of the network is very high. A two-stage joint routing approach was suggested in [16]. Muhammad Asshad et al. [17] investigate the utility of Rayleigh and Weibull fading channels in non-regenerative wireless cooperative networks using the strongest relay selection methodology. The Moment Generation Function (MGF) of the signal-to-noise relationship (SNR) at the expected node was determined using a Weibull fading parameter. Calculate the lower bound error symbol value and the probability of failure using MGF. The relay selection technique is used to check the precision of a derivation in a variable number of relay nodules based on the study and simulation results on the likelihood of failure and the symbol error rate. Nonetheless, due to the accuracy of the analytical model and the waste of resources, there are far less relay nodes.

By converting the original problem into the analogue finite-horizon Markov Decision Process (MDP) with fixed-stages numbers Bojie LV et al. [18] developed asymptotically an optimized solution system. In the analysis of the approximation functions, a new approach method was developed to solve the dimensionality curse. There are also analytical constraints on the exact value function and the approximation error. The value functions estimated will depend on certain device information, e.g. distribution request. A reinforcement learning algorithm was suggested for a situation where these numbers are unclear. In this research work, the use of energy is also a key topic.

Yuan Gao et al. [19] proposed a novel system of selection and fusion transmission of nodes using artificial intelligence: initially, draw the status of the mobile terminals that represent the thermal pattern, and then suggest a deep learning approach to indicate the status of each node and allow an optimized selection of the target node; finally, perform a multi-stage wireless information fusion transfer to increase spectrum quality. In addition, as many users use the same data, the core network and the wireless connection can have high pressure.

Elhawary et al. [20] nearest nodes are hired for communication assistance in shared networks by transmitting and receiving nodes. Cooperative wireless network communication links are configured as cluster transmitters and receivers. A co-operative exchange protocol was then suggested for the development and co-operative transmission of data for these clusters. The upper limit of protocol capabilities was obtained and the final robustness and trade-offs between the energy consumption and the error rate of the data packet loss protocol were evaluated.

Although Ullah et al. [21] affects the accuracy of the analytical model due to a large reduction in the number of relay nodes, Wang et al. [22] the main problems are routing and energy consumption, Singh and Singh [23] the core network and the wireless link will provide high pressure when multiple users use the same data, Tariq et al. [24] does not control the packet error rate (PER), maximizing the confidentiality rate, Gao et al. [25] obtains a poor solution and low convergence. From the above problems, it is important to create a new methodology for routing energy consumption algorithms in a cooperative network.

3

COOPERATIVE ROUTING ALGORITHM FOR LARGE-SCALE WIRELESS NETWORKS

3.1 Introduction to Cooperative Relay Networks

Cooperative relay networks have become an effective tool in wireless access networks to fight multipath fading and rising energy consumption [26]. A diversity study of single and multiple relay selection schemes was performed to generalize the concept of relay selection by encouraging more than one relay to cooperate. In simulations, multiple relay selection methods performed much better than single relay selection methods [27]. As per the analysis of single and multiple relay selection for the cooperative communication model, selecting more than three relays yielded a marginal gain. The effectiveness of a decentralized relay selection algorithm [28], where each relay node determines whether it should relay or stay silent depending on its own instant channel gain and a predetermined threshold was compared to the opportunistic relay selection method [29]. The results of network geometry by constructing interacting groups of terminals as a result of relay selection is one of the directions illuminated in [30]. Yang et al. [31] considered a cooperative system for multi-antenna terminals in which the source agrees to join only when it requires assistance based on a fixed threshold. The source only chooses one relay based on the maximum instantaneous value of the relay's metric. Many AF relays share an orthogonal channel with the source, showing optimal Diversity-Multiplexing Trade-off (DMT) [32]. Cooperative diversity based on STBC seems to have a stronger

DOI: 10.1201/9781032714011-3

diversity order than repetition-based algorithms, and it can be used to improve spectral performance.

In cooperative communication networks, relay selection methods have been used to improve wireless system performance in a variety of ways, including throughput, channel bandwidth, failure probability, and coverage extension [33]. The SNR of the receiver is used to determine a high-performance wireless system, which is then connected to the required power level at the receiving node. The obtained SNR using either AF or DF is used to quantify the capacity of relaying schemes in cooperative networks, as seen in equation 3.1 [34]. Equation 3.2 also measures the relationship between the feasible transmission rate and the transmit power [35].

$$C_{DF} = \log\left(1 + \min\left\{\gamma_1, \gamma_2, \ldots\ldots, \gamma_M\right\}\right)$$
$$C_{AF} = \log\left(1 + \gamma_{total}\right) \tag{3.1}$$

$$R = W \log\left(1 + \frac{P}{WN_0}\right) \tag{3.2}$$

Under a power constraint, relay selection strategies were investigated from various performance perspectives. Some authors [36,37] proposed using a one-bit input to approximate the relation SNR and a predefined threshold to choose the relay when power is restricted. The proposed relay selection technique in [38] uses the SNR threshold at the first hop to save power by shutting off relays with SNR below the threshold when the outage risk is heavy. In [39], authors have been proposed an end-to-end SNR level to limit network capacity consumption while throughput is limited. When data rates are limited, Dan et al. [40] suggest a mechanism for reducing energy usage. In this process, the energy level of applicant relays was factored into the selection metric, extending the total network life span. When there is a time limit, power saving is activated in [41]. Furthermore, for cooperative networks, energy-efficient relay selection schemes have been proposed using various selection parameters like minimizing the SNR at the destination and selecting the hop with the best channel condition [42,43]. Relay selection schemes and their effect on increased energy expense per transmission have also been activated in [44,45] from the perspective of battery life. Furthermore, Long et al. [38] suggest relay selection based on Channel State Information (CSI), while Marques et al. [46] suggest using finite CSI as input to the transmit cluster to choose the right channel condition node.

To improve energy consumption, a minimal distance relay selection process [47] was used. In addition, to reduce Symbol Error Probability, a relay selection strategy focused on relay position has been proposed in [48]. Similarly, the authors of [130] suggest minimizing transmit power when aiming for a mean square error target, while the authors of [49] propose allocating transmit power dependent on BER specifications. Relay selection based on optimal propagation distance and node residual resources, on the other side, a routing technique to extend network lifetime has been presented in [50]. The relay selection approach in [51] uses simulated MIMO transmission between clusters to optimize the SNR. Furthermore, the impact of relay location on energy efficiency was investigated in [52] and single and multiple relay selection methods centered on relay distance from the source and endpoint were proposed. Similarly, some authors [53,54] looked at the effect of relay position on energy efficiency from the standpoint of power minimization and equal capacity allocation among nodes. The relay channel implementation mitigates the negative effect of route failure in cooperative communication since the transmission topology is focused on shorter connections. Because of the number of pathways accessible with varying lengths and channel requirements, the usage of more than a few nodes in a cooperative network is unavoidable in order to provide better transmission consistency, which increases the wireless system's energy performance.

However, in order to preserve high efficiency, relay selection ought to be paired with device optimization parameters to ensure energy-efficient cooperative communication. As a consequence, the channel state, modulation technique should be considered in the optimization policy. We suggest a low-energy single relay selection algorithm that takes advantage of the transmitted aspect of the wireless medium by sending the source signal to the node with the shortest two route lengths. We approach power savings by choosing relays that offer the shortest total transmission route while also fulfilling device efficiency criteria, as the proposed approaches are based on lowering the energy cost per bit. We often use numerical analysis to examine the influence of modulation strategy on power consumption and suggest the best approach. Furthermore, we examine both coded and un-coded processes to demonstrate how they impact energy use. As a consequence, with various propagation lengths, different channels are considered. We use a similar definition to [55], except the energy cost is determined depending on the transmitting range duration and a targeted BER at the destination.

Energy performance, or energy conservation, is another significant parameter for cooperative routing (CR) algorithms. Li et al. [56] suggested a low-complexity algorithm for energy-efficient CC, as well as optimal power allocation, channel allocation, and relay collection. To minimize network energy usage, Akhtar et al. [57] introduced a protocol that optimizes the MAC, physical layer, and network layer efficiency all at once. A cooperative automatic repeat request (ARQ) scheme was established at the MAC layer, and cooperative cost-based shortest path routing (CCB-SPR) and cooperation over non-cooperative shortest path routing were developed at the network layer (CONSP).

Another measure of stability is the network life cycle; Zhang et al. [58] proposed a routing algorithm that uses virtual MIMO to improve node efficiency by extending the life of the node involved. A CR algorithm combining cooperative plurality, distributed energy-aware routing strategy, and truncated ARQ scheme has been proposed to maximize network life cycle. El Monser et al. [59] introduced low energy adaptive clustering hierarchy distributed cooperative relaying, a low energy consumption protocol based on the CC methodology. The authors consider a network of single radio terminal (Omni-directional antenna) nodes; as a consequence, signals can transmit in an unwanted transmitting direction, lowering the efficiency of the node.

Habibi et al. [60] developed a mathematical model for relay assignment based on branch and bound framework for multi-hop cooperative networks with single radio terminal. The computational complexity of the mathematical approach's solution grows exponentially. To increase throughput, the workload aware opportunistic routing algorithm was proposed. This assumption does not improve the throughput of multiple flows because it assigns the same channel to all nodes that are part of the flow. Aung and Chong [61] proposed a cluster-based multi hop CR with a distributed agent-based coordination (DAC) mechanism. According to the authors' simulation, this mechanism in the MRMC network improves throughput by about 200 percent over a single radio single channel network. However, the authors assumed that each node in a cluster uses the same reception channel.

Mostafaei [62] proposed a reliable routing to improve reliability by sending packets over the best path with the highest packet delivery ratio, which is determined using a learning algorithm. To improve reliability, Kim et al. [63] proposed an opportunistic routing algorithm. Every node in this scheme is capable of selecting a relay node and dynamically handling link failures for each transmission. To extend the network lifetime, a novel compressed sensing approach was presented. The authors also proposed a method for determining the best basis representation using a machine learning algorithm. For multi-hop cooperative networks, Gao et al. [15] proposed an opportunistic cooperative packet transmission (OCPT) scheme. A cluster head selects the transmitter and receivers to form MIMO in OCPT before the transmission. The number of transmitters and receivers in each hop increases the end-to-end transmission time and energy consumption per node.

Bai et al. [64] proposed constructive relay-based cooperative routing (CRCPR), a system for preventing link failures due to mobility. For a mission-critical push-to-talk (MCPTT) system, a routing strategy based on link reliability and transmission delay was presented. A routing approach based on integrating reinforcement learning algorithms was presented to address routing protocol problems caused by dynamism in network topology. Venkatesh et al. [65] proposed a two-hop geographic opportunistic routing (THGOR) that selects a set of nodes with a high packet delivery ratio to improve energy efficiency and packet delivery ratio. Saravanan [66] proposed an energy-efficient routing protocol to extend the lifetime of a network without sacrificing capacity. The use of modified ant colony optimization was used to present a shortest path routing algorithm that selects the best path from all possible routes (ACO). Despite the fact that multi-radio CR algorithms provide better performance, they result in a significant transmission delay.

In large-scale CC, determining the shortest path routing with candidate CA and minimal sessions is more important than determining the shortest path routing with candidate CA alone. In light of the foregoing, this thesis proposes a dynamic routing algorithm that considers the following parameters along the line of sight (LOS) between source and destination:

- *Maximum available channel capacity (MACC):* A node calculates the available channel capacity (ACC) between it and each node in its transmitting coverage area. Since high throughput necessitates high channel capacity, the link with the highest MACC is chosen as the candidate channel, with the corresponding node acting as a relay.
- *Distance (DS) parameter:* A node nearest to the LOS is used as the relay in order to minimize the number of hops.

From the literature it was observed that the joint optimization of routing – CA and relay selections – cooperative routing is NP. To solve the problem we propose a solution by considering four parameters, i.e. relay selection, routing, CA and end-to-end transmission time with two important components: cooperative relay selection algorithm and delay reduction algorithm as illustrated in Figure 3.1.

To begin, every node calculates the MACC routing metric on a regular basis in order to achieve high network throughput and minimize co-channel interference. When a new flow comes, a cooperative relay selection routing algorithm is distributed to find the cooperative routing route based on this metric.

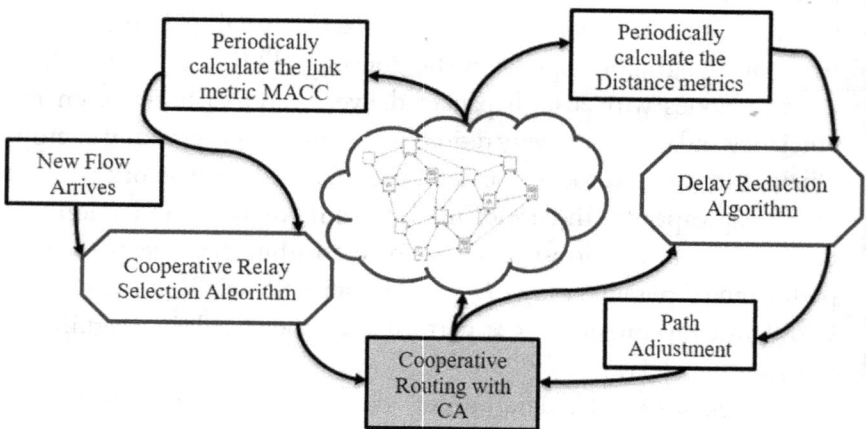

Figure 3.1 Solution framework.

Second, to obtain the routing path with minimal number of hops, delay reduction routing algorithm is run based on distance parameters, and the routing path and channel adjustment will be done accordingly.

3.2 System Model

As seen in Figure 3.2, we assume a multi-hop multi-channel cooperative wireless network. There are N nodes, each of which are provided with multiple radio terminals (multi radio) and are spread over a region of $LxLm^2$. Mobility is believed to have a marginal effect. The Amplify and Forward (AF) relaying protocol is used by each node [67]. With M orthogonal channels $CH = \left\{ ch_1, ch_2, \ldots\ldots\ldots\ldots, ch_M \right\}$ and K concurrent flows throughout the network, each node is presumed to have the same transmitted power P_t. Any node uses GPS to know its geographical position, and other nodes submit periodic beacon signals to determine their location [68]. Table 3.1 lists the notations used in this chapter.

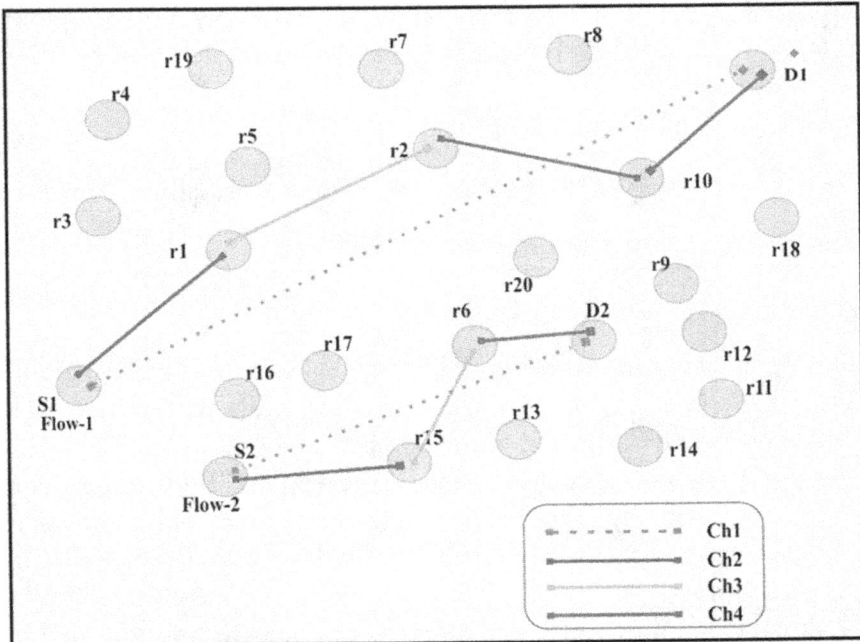

Figure 3.2 Multi-radio-multi-channel cooperative network.

Table 3.1 Summary of notations

SYMBOL	DESCRIPTION
s	Source Node
d	Destination Node
N	Number of Nodes
LxL	Network Area
P_t	Transmitted Power
K	Number of Concurrent Flows
M	Number of Orthogonal Channels
CH	Set of Orthogonal Channels
ch_i	i^{th} Orthogonal Channel
B	Channel Band Width
h_{ij}	Channel Coefficient on the link $i \rightarrow j$
A	Amplification Factor
η_{ij}	Additive White Gaussian Noise between nodes i and j
SNR_{m,n,ch_i}	Signal to Noise Ratio of channel ch_i between node m and n
C_{m,n,ch_i}	Total Capacity on Channel ch_i between node m and n
I_m	Set of nodes in the interference region of node m
L_{p,q,ch_i}	Load on channel chi between node p and node q
$A_c(s)$	Set of nodes under the coverage area of node s which is evaluated based on predefined threshold
τ_1	SER threshold
ϕ	Null Set
$LOS_{s,d}$	Line of Sight from node s to node d
$ch_{ca}^{s,j}$	Candidate Channel between node s and node j
$MACC_{s,j;ch_i}$	Set of m+aximum available channel capacity between node s and node j
α	Capacity Threshold
β	Distance Threshold
ς	Set of working channels in co-channel interference region
ψ	set of nodes which are having relative capacity difference less than capacity threshold
$w(j)$	set of working channels of node j

We assume that the network has a global table at the access point that each node in the network may access. This table stores each node's working channels, allowing each node to be aware of the working channels of all other nodes in the network. The configuration channel is used by each node to upgrade its functioning channels. This table comes in handy during the CA process, as discussed in Algorithm 1.

Figure 3.3 illustrates a single hop along the path from source s to destination d. As seen in equation 3.3, $A_c(s)$ is the set of nodes under

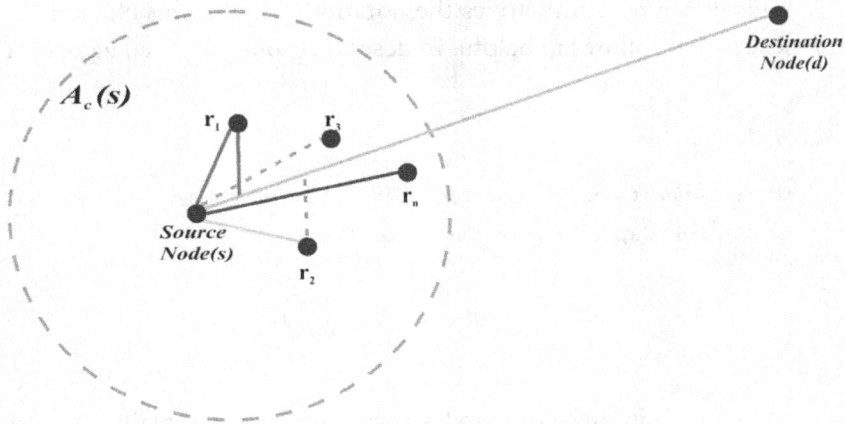

Figure 3.3 Hop of the route between s and d.

the coverage region of node s where the symbol error rate (SER) is less than a predefined threshold τ_1.

$$A_c(s) = \left\{r \middle| SER_{s,r_i} < \tau_1\right\} \tag{3.3}$$

Based on MACC and DS parameters, source node (s) selects one of the nodes r_j ($r_j \in A_c(s)$) as relay node. For the next hop, the chosen relay node becomes the source node. The relay node collects data x from the source node in each hop and amplifies it before re-transmitting it to the relay node in the next hop. The received signal at the node j from node i is given as [69],

$$y_{ij} = \sqrt{P_t}\, h_{ij} x + \eta_{ij} \tag{3.4}$$

This data is amplified at node j and re-transmitted to the next hop relay l. The transmitted data is given as:

$$y_{jl} = A\sqrt{P_t}\, h_{jl} y_{ij} + \eta_{jl} \tag{3.5}$$

3.3 Routing Algorithm

To increase network throughput and reduce propagation time, the CR algorithm and delay reduction (DR) algorithms are suggested in this chapter. Table 3.1 summarizes the notations used in this chapter. The concepts that follow are helpful in describing our proposed algorithm.

3.3.1 Definitions

- *Available Channel Capacity (ACC):* It specifies a channel's available capacity. It is graded as follows:

$$ACC_{m,n;ch_i} = C_{m,n;ch_i} - \sum_{p,q \in I_m} L_{p,q;ch_i} \qquad (3.6)$$

where I_m is the set of nodes in the interference region of node m, $L_{p,q;ch_i}$ is the traffic load on the channel ch_i between the nodes p and q; $C_{m,n;ch_i}$ is the total capacity of channel chi between the nodes m and n, $C_{m,n;ch_i}$ can be calculated as:

$$C_{m,n;ch_i} = B.\log\left(1 + SNR_{m,n;ch_i}\right) \qquad (3.7)$$

- *Candidate channel (ch_{ca}):* A channel over which two nodes communicate and having maximum ACC. The candidate channel $ch_{ca}^{m,n}$ between nodes m and n is identified as:

$$ch_{ca}^{m,n} = \max\left\{ ACC_{m,n;ch_i} \left| ch_i \in CH \right. \right\} \qquad (3.8)$$

3.3.2 Cooperative Relay Selection Algorithm

We proposed a MACC-based relay selection algorithm in this segment, which determines the maximum ability route between the source and destination. Relay selection and CA are defined in depth in Algorithm 1. The dynamic cooperative relay discovery algorithm is used to identify the relay nodes, channels, and routing path when a new flow request arrives in the network. Each flow's source node runs

Algorithm 1 on a regular basis and updates the route. This algorithm only considers the MACC parameter, which could result in a long transmission delay.

Algorithm 1 Cooperative Relay Selection Algorithm

Input: A flow request from node s to node d
Output: Routing path from s to d with each hop's relay node, and its channel

1: **while** source node $\neq$ destination node do
2: Find $A_c(s)$
3: Source node s calculates the set of working channels ς in its co-channel interference region (I_s) using global table.

$$\varsigma = \bigcup_{j \in I_s} w(j)$$

and evaluates the complements $\varsigma' = CH - \varsigma$
4: It forms a subset of $A_c(s)$, which lies between source and destination.

$$DD(s) = \{j | D(j,d) < D(s,d), j \in A_c(s)\}$$

5: **if** $d \in A_c(s)$ **then**
6: **if** $\varsigma' \neq \phi$ **then**
7: Find the candidate channel between source and destination nodes

8: $ch_{ca}^{s,d} = \max \left\{ ACC_{s,d;ch_i} \mid ch_i \in \varsigma' \right\}$

9: **else**
10: Find the candidate channel between source and destination nodes

11: $ch_{ca}^{s,d} = \max \left\{ ACC_{s,d;ch_i} \mid ch_i \in CH \right\}$

12: **end if**
13: **else**
14: **if** $\varsigma' \neq \phi$ **then**
15: Find the candidate channel between source and every node j

16: $ch_{ca}^{s,j} = \max \left\{ ACC_{s,j;ch_i} \mid ch_i \in \varsigma', j \in DD(s) \right\}$

17: **else**
18: Find the candidate channel between source and every node j

19: $ch_{ca}^{s,j} = \max \left\{ ACC_{s,j;ch_i} \mid ch_i \in CH, j \in DD(s) \right\}$

20: **end if**
21: **end if**
22: The node which is having MACC is selected as relay node.

23: $relay\ node = \left\{ j \mid \max \left(ACC_{s,j;ch_i} \right) \right\}$

24: Assign the candidate channel for the pair: source node s and relay node j i.e., link $s \rightarrow j$.
25: source node = relay node
26: **end while**

3.3.3 DR Algorithm

A DR Algorithm is suggested that considers the distance parameter in addition to the MACC parameter to improve transmission delay.

Figures 3.4 and 3.5 depict a DR procedure in action. Allow data to flow from node A to node G. The routing direction provided by Algorithm 1 for this flow is $A \rightarrow B \rightarrow E \rightarrow F \rightarrow G$. If there is a node C that is closer to the LOS and $ACC_{A,C;ch_3}$ is nearly identical to $ACC_{A,B;ch_1}$, as seen in Figure 3.5, the relay node is then moved from node B to node C by the DR algorithm. As a consequence, the DR algorithm's resultant direction is $A \rightarrow C \rightarrow F \rightarrow G$. Below is a summary of the DR algorithm.

Algorithm 2 Delay Reduction Algorithm

Input: ACC of link between node s and each node in $A_c(s)$ and its candidate channel.
Output: Updated routing path between s and d with less number of hops

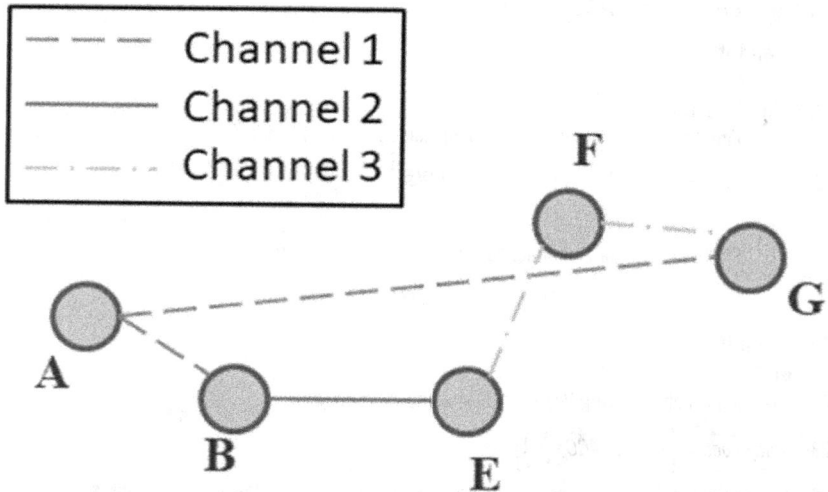

Figure 3.4 An illustration of a CR algorithm.

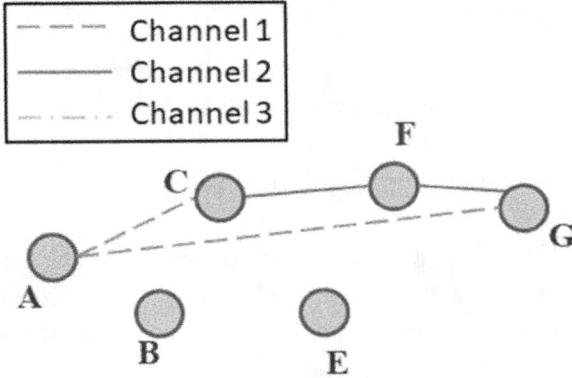

Figure 3.5 An illustration of DR algorithm.

1: **while** source node ≠ destination node **do**

2: Form a set $MACC_{s,j;ch_j} = \left\{ ACC_{s,j;ch_{ca}} \mid j \in DD(s) \right\}$

3: Form a set of nodes ψ which satisfy the condition,

$$\psi = \left\{ z \mid \max\left(MACC_{s,j;ch_{ca}} \right) - MACC_{s,j;ch_{ca}} < \alpha; z, j \in DD(s) \right\}$$

where α is a predefined capacity threshold

4: **if** length of ψ == 1 **then**

5: relay node = z

6: else

7: Form a set of perpendicular distances from a node $j(j \in \psi)$ to LOS

$$\chi_{Z, LOS} = \{ D(Z, LOS) \mid Z \in \psi \}$$

8: Form a set of nodes which satisfy the condition

$$\xi = \left\{ y \mid \chi_{y,LOS} - \min\left(\chi_{y,LOS} \right) < \beta; y \in \psi \right\}$$

where β is predefined distance threshold

9: **if** length of ξ == 1 **then**

10: *relay node = y*

11: **else**

12: *relay node* $= \{ j \mid D(s, j) > D(s, j),\ i \neq j, j \in \xi \}$

13: **end if**

14: **end if**

15: Assign the candidate channel for the pair: source node s and relay node j i.e., link $s \rightarrow j$.

16: source node = relay node

17: **end while**

3.3.4 Summary of DR Algorithm

1. Source node specifies the range of nodes accessible in its transmission coverage area, as well as their working channels using the global table.
2. It is a subset of step 1, and it satisfies the direction deviate metric.
3. Using Equation 3.6, the source node seeks a candidate channel from the accessible free channels to any node in its radius.
4. It produces a group of nodes with a relative capacity gap less than a predetermined threshold.
5. It is a subclass of step 4 that has a relative distance discrepancy less than or equal to a predetermined distance threshold.
6. A relay node far from the source is chosen from the collection of nodes in step 5. This relay node serves as the next hop's root node.
7. For contact between source node and relay node, a candidate channel is allocated, and relay node will serve as source node for the next hop.

This cycle is repeated before the relay node is designated as the destination. Figure 3.6 depicts the DR algorithm's flow diagram. The flow chart replicates the DR algorithm and the same repeated until the optimal result is achieved.

3.4 Simulation Results

The simulation results of our proposed algorithms are presented in this portion, along with a comparison to the performance of the IACR algorithm. 75 nodes are randomly generated in a $1,000 \times 1,000$ m^2 area to carry out the simulations (unless otherwise specified). Using MATLAB® and the simulation parameters specified in Table 3.2, we evaluated the efficiency of our proposed algorithms.

As seen in Figure 3.7, the suggested DR algorithm is simulated for three concurrent flows: flow-1 (node 1 to node 15), flow-2 (node 18 to node 69), and flow-3 (node 5 to node 39). Below are the routing paths for all three flows (along with the CA):

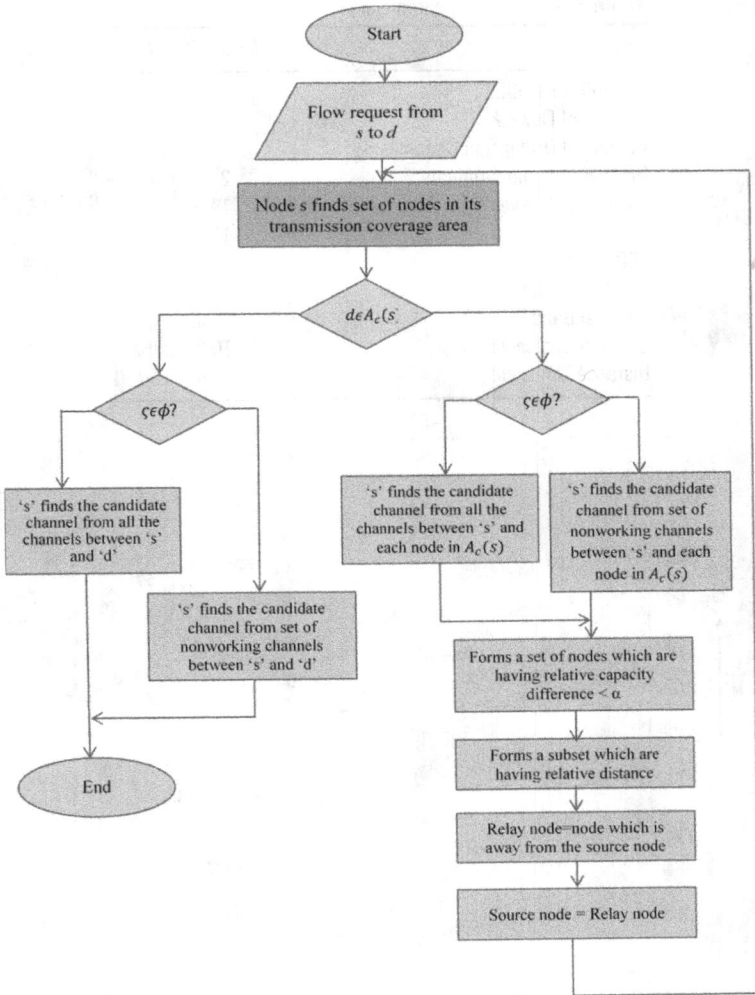

Figure 3.6 Flow diagram of the DR algorithm.

$$flow\ 1: 1 \overset{ch_4}{\Rightarrow} 62 \overset{ch_5}{\Rightarrow} 6 \overset{ch_4}{\Rightarrow} 60 \overset{ch_5}{\Rightarrow} 33 \overset{ch_4}{\Rightarrow} 53 \overset{ch_1}{\Rightarrow} 15$$

$$flow\ 2: 18 \overset{ch_3}{\Rightarrow} 6 \overset{ch_1}{\Rightarrow} 60 \overset{ch_3}{\Rightarrow} 4 \overset{ch_1}{\Rightarrow} 1$$

$$flow\ 3: 5 \overset{ch_3}{\Rightarrow} 73 \overset{ch_2}{\Rightarrow} 66 \overset{ch_5}{\Rightarrow} 35 \overset{ch_3}{\Rightarrow} 69 \overset{ch_2}{\Rightarrow} 39$$

Table 3.2 Simulation parameters

SYMBOL	DESCRIPTION
NUMBER OF NODES	75
Number of Flows K	3
Number of Orthogonal Channels M	5
Number of Radio Terminals	2
Transmitted Power P_t	1W
Noise Variance	10^{-10}
SER Threshold τ_1	10^{-3}
Interference Range	200m
Channel Band Width	22MHz
Capacity Threshold α	1000bps/Hz
Distance Threshold β	25m

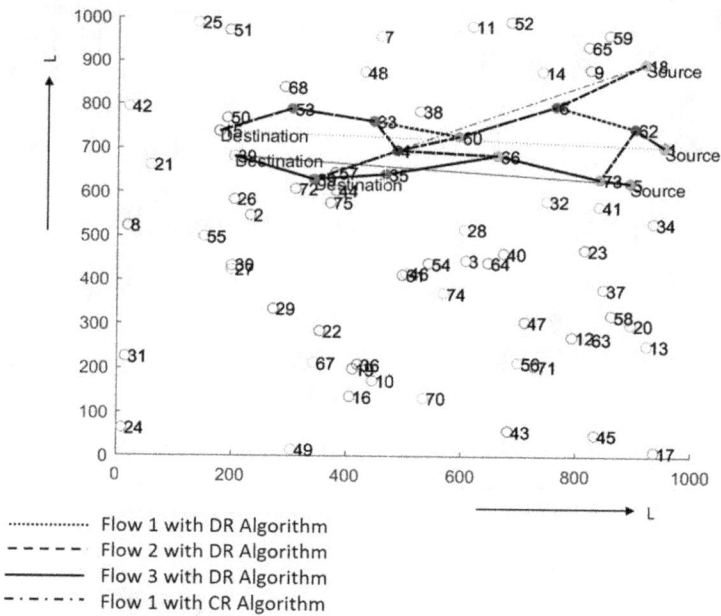

.... Flow 1 with DR Algorithm
– – – – – Flow 2 with DR Algorithm
————— Flow 3 with DR Algorithm
– · – · – ·· Flow 1 with CR Algorithm

Figure 3.7 Path of concurrent flows.

The CR algorithm's routing path for flow-1 is:

$$flow\,1 : 1 \overset{ch_4}{\Rightarrow} 62 \overset{ch_5}{\Rightarrow} 73 \overset{ch_4}{\Rightarrow} 66 \overset{ch_5}{\Rightarrow} 4 \overset{ch_5}{\Rightarrow} 33 \overset{ch_4}{\Rightarrow} 53 \overset{ch_1}{\Rightarrow} 15$$

Figure 3.7 depicts the routing paths for the three flows described above. It can be shown that nodes 6 and 60 are involved in two separate flows, flow-1 and flow-2. Since each node has several radios, it will participate in several flows and interact through multiple orthogonal channels to prevent interfering with each other (depends on CA). The proposed DR algorithm assigns channel ch_4 to connection 6 => 60 in flow-1 and channel ch_1 to the same link in flow-2, thus eliminating interference.

By updating the global table at the entry point, the channel calculation maximizes the gain. The nodes attached to the candidate channels can be dynamically determined, and the data can be used to change the global table to show the node is currently accessing the working channel. Table 3.3 depicts an instance of the global table when the three concurrent flows described earlier are served throughout the network (note that Table 3.3 lists out information about the nodes which are involved in those three flows). In Table 3.3, operating

Table 3.3 Global table

NODE NUMBER	CH$_1$	CH$_2$	CH$_3$	CH$_4$	CH$_5$
1					
4					
5					
6					
15					
18					
33					
35					
39					
53					
60					
62					
66					
69					
73					

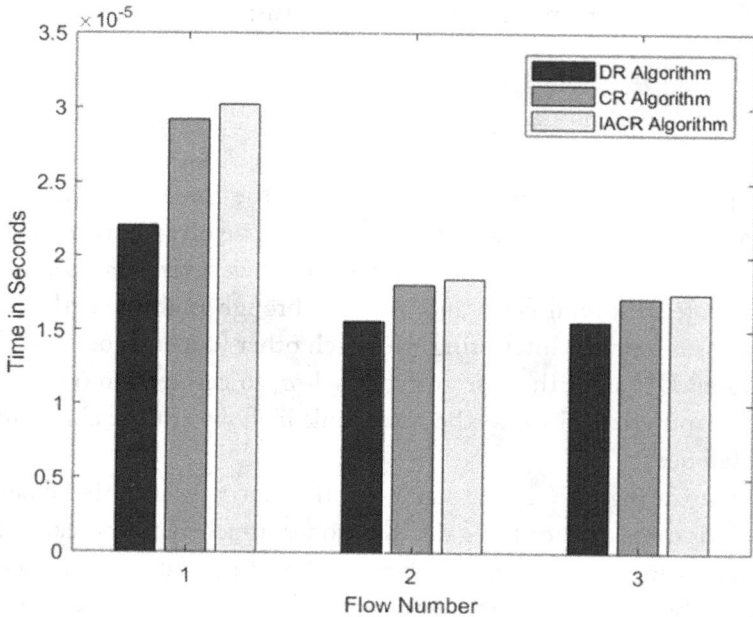

Figure 3.8 Transmission delay.

channels are denoted by a red block, while idle channels are denoted by an empty node. Nodes 4 uses channels 1 and 3 and is thus depicted in red, whereas the other channels are not utilized for this node and are thus left empty.

The transmission times of the three flows evaluated for the CR, DR, and IACR algorithms are shown in Figure 3.8. Since the relay node in IACR and CR algorithms is only chosen based on the ACC, the routing path can deviate from the LOS, increasing the number of hops, as seen in Figure 3.9. The DR algorithm chooses a node as a relay node if it follows the following criteria:

1. Away from the source node
2. Closer to the LOS.

Since a node farther out from the source node ensures a minimal number of hops, and a node closer to the LOS ensures the shortest distance, resulting in the shortest transmission delay. Since the number of hops is reduced, the re-transmission delay of relay nodes is reduced, and the transmission time for the DR algorithm is reduced.

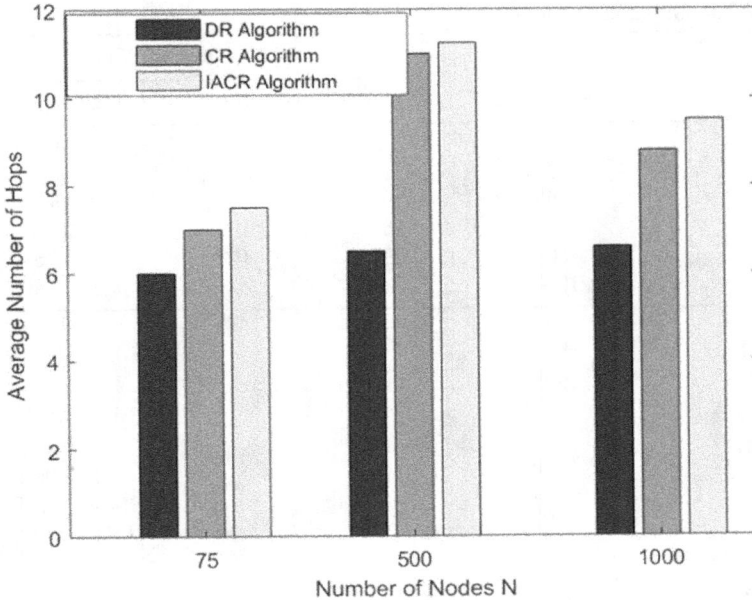

Figure 3.9 Average number of hops over number of nodes N, $L = 1,000$ m.

The transmission time of flow 1 for IACR algorithm is 3.1×10^{-5} sec and for DR algorithm is only 2.2×10^{-5} sec, an improvement of 27.5 percent.

If the network area expands, the distance between nodes grows, and the total number of hops continues to grow as well. Table 3.4 shows the total number of hops for various algorithms for various L and N. When L = 5,000 m, the IACR algorithm's total number of hops is about three times that of the DR algorithm.

Figure 3.10 shows how the number of orthogonal channels affects the network's efficiency. We calculated aggregate throughput by varying the number of channels from 1 to 20, while holding all other parameters at their default values (see Table 3.2). The aggregate throughput increases initially with the number of channels and then stays stable until the number of channels is adequate for the three flows, according to the simulation results. Extra channels are not utilized by the network since each node has only two radio terminals and the number of flows considered is three.

Table 3.4 Average number of hops

		AVERAGE NUMBER OF HOPS		
L IN M	NUMBER OF NODES	IACR	CR ALGORITHM	DR ALGORITHM
1000	75	7.5	7	6
	500	11.25	11	6.5
	1000	9.5	8.8	6.6
5000	75	25.3	21.7	16
	500	39.7	34.8	17.5
	1000	28.5	26	17.8

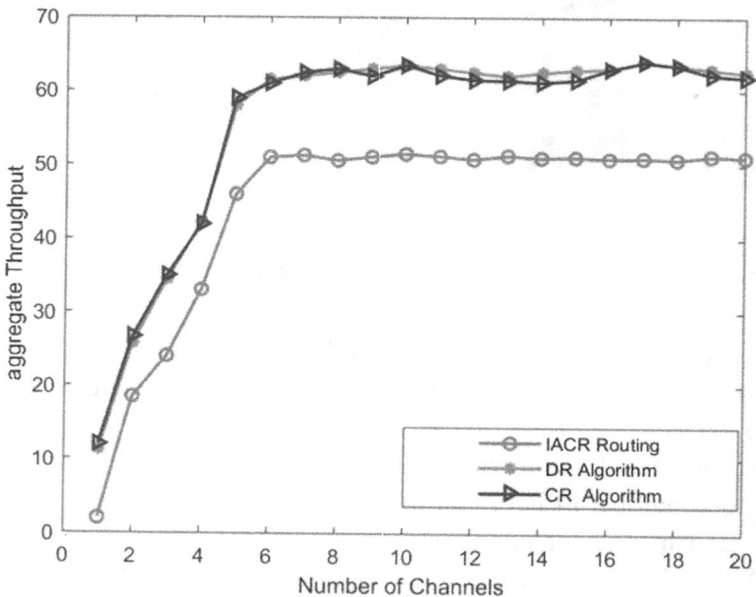

Figure 3.10 Impact of number of channels.

3.5 Summary

Two complex routing algorithms focused on MACC and distance parameters are introduced in this chapter to optimize the advantages of CR in the multi-radio-multi-channel large-scale wireless network. The suggested algorithms' combined throughput and transmission time was equivalent to the IACR algorithm [70]. The best relay node among candidate relay nodes with better channel capability is found by using nodes close to LOS between source and destination and away from the source node as a guide. At the access point, the complex CA with high channel capability is formulated using the global chart. According to the simulation, the suggested algorithm improves propagation latency and average number of hops by 27.5 and 32.5 percent, respectively.

4

ENERGY-EFFICIENT TRANSMISSION FOR LARGE-SCALE COOPERATIVE WIRELESS NETWORKS

4.1 Introduction

Communication between two mobile nodes is triggered by a base station (BS) in Mobile Ad Hoc Networks (MANET), and the two nodes will then connect with one another without involving the BS. In the case that the link between these two nodes is weak (poor), intermediate mobile nodes (relays) work with the source node to transfer data to the destination. The data is obtained from different directions (known as spatial diversity) at the destination node, resulting in improved efficiency. Cooperative communication is the term for this situation (CC). To take advantage of the benefits of MIMO, the CC technique is now used in nearly all wireless networks. Since the position of a relay affects the efficiency of wireless networks, the majority of research has focused on relay placement, and it has been discovered that the relay that is in the middle of the network performs best [71]. To boost network efficiency, CC can be applied to large-scale networks. To increase the lifecycle of mobile nodes/networks, network energy consumption becomes a crucial consideration when choosing cooperative relay nodes (routing). By incorporating the idea of a super node, Akhtar and Nakhai [72] suggested a power conscious cooperative routing algorithm.

To increase throughput and reduce total path failure, cooperative relay networks are being used. Cooperative transmission is used to

                          DOI: 10.1201/9781032714011-4

increase reliability and spatial diversity. Amplify and Forward (AF), Compress and Forward (CF), and Decode and Forward (DF) are three relaying protocols that cognitive relay networks may use. A relaying protocol is widely used to create a reliable communication link between the two nodes when direct connectivity between them is not feasible or the quality of the channel between them is limited. To maximize throughput and the SNR, relay nodes are placed between the destination and source nodes. Throughout the single hop relaying protocol, the single relay is positioned between both the destination and source nodes. The multi-hop relaying protocol has been used where more than one relay is positioned between the destination and source nodes.

The authors of [73] suggested an efficient low energy routing by weighting the route with power rate ratios. A routing algorithm based on the number of transmitting and receiving antennas and a collection of cooperative nodes has been suggested in [74] to achieve the lowest energy consumption. Praveen Kumar et al. [75] suggest an energy-balanced cooperative routing algorithm. The authors propose a MIMO-based routing algorithm to reduce energy consumption by distributing power evenly among transmitters. In [76], a cooperative ARQ architecture was developed at the MAC layer to reduce total network energy usage, and an ARQ-based energy efficient routing algorithm was presented. Pandana et al. [77] propose distributed cooperative routing algorithms based on maximum power distribution. With end-to-end throughput restrictions, Zheng et al. [78] suggested an energy-efficient cooperative routing with truncated ARQ to maximize overall energy consumption.

Many of the authors above have suggested different energy-efficient ways to minimize energy demand, but none of them are applicable to multiple radio nodes. In this chapter, we suggest an Energy Efficient Transmission (EET) scheme for large-scale wireless networks with multiple radios and multiple hops.

To increase spectral accuracy, Rankov and Wittneben [79] developed different DF and AF cooperative relaying protocols. Han et al. [80] found that perhaps the spectral performance loss of the two-way AF half-duplex relay has near lower and upper limits. Peng et al. [81] measured the performance frequency of the two-way cognitive relay network using the DF protocol via Rayleigh fading channels.

With relaying protocol, channel models and relay collaboration, Naeem and Rehmani [82] go into the basics of a cognitive radio network. Cognitive relay networks with decode and forward relay were designed by Xu et al. [83]. They increased the SNR and measured the probability of a failure. Several studies have suggested AF and DF over Rayleigh fading and Nagakami fading sources with cooperative spectrum sensing systems. The possibility of symbol error in DF over Rayleigh fading channels was also explored.

In order to achieve efficiency over Nakagami and Rayleigh fading channels, a variety of studies use single hop in which some form of relay is being used in the cognitive relay system. In several of the experiments, multiple relays have been used to counteract Rayleigh's fading. Using only a multi-hop relay system, the reliability of spectrum utilization can be enhanced, user latency can be minimized, network capacity can indeed be expanded, and network coverage could be expanded.

4.2 Cooperative Communication Over Single-Hop Networks

There has been direct connection between source (S) and destination (D), and a cooperative link through a relay [9,84]. As a result, the system is actually called a single hop relay system, as shown in Figure 4.1.

A half-duplex downlink system is similar to this example. In single hop cooperative communication, information is transmitted in two phases. In phase I, source broadcasts the information which can be received by the destination and relay. The information at the destination and relay in phase I can be expressed as:

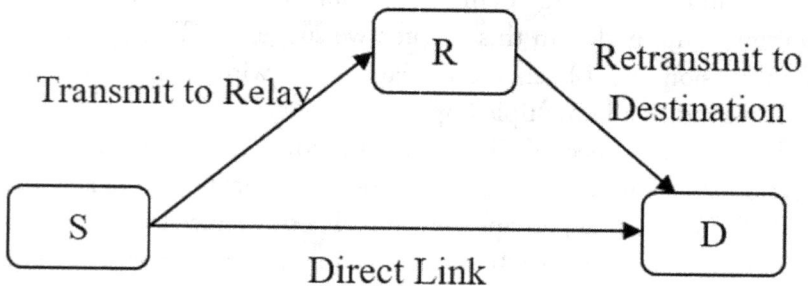

Figure 4.1 Relay scheme with a single hop.

$$Y_{SD} = x h_{SD} \sqrt{P_t} + \eta_{SD} \tag{4.1}$$

$$Y_{SR} = x h_{SR} \sqrt{P_t} + \eta_{SR} \tag{4.2}$$

h_{SD} and h_{SR} are the channel coefficients between source-to-destination and source-to-relay respectively; and P_t is the transmitted power, x information being transmitted by the source. η_{SD} and η_{SR} are the additive white Gaussian noise between the links source-to-destination and source-to-relay with zero mean. The instantaneous signal to noise ratio (SNR) of a source-to-relay link can be given by:

$$\gamma_{SR} = \frac{\left|h_{SR}\right|^2 P_t}{N_0} \tag{4.3}$$

Where N_0 is referred to as the noise power.

In phase II (also called cooperative phase), relay node amplifies the received information from the source and re-transmit to the destination. The information at the destination in cooperative phase is given by:

$$Y_{RD} = A \sqrt{P_r} h_{RD} Y_{SR} + \eta_{RD} \tag{4.4}$$

$$Y_{RD} = A \sqrt{P_r} \sqrt{P_t} h_{SR} h_{RD} x + A \sqrt{P_r} h_{RD} \eta_{SR} + \eta_{RD} \tag{4.5}$$

Where A is the amplification factor can be expressed as:

$$A = \sqrt{\frac{P_r}{P_t \left|h_{SR}\right|^2 + N_0}} \tag{4.6}$$

The effective noise term in equation 4.5 is equivalent to the noise variance N_0, i.e.,

$$N_0 = A \sqrt{P_r} h_{RD} \eta_{SR} + \eta_{RD} \tag{4.7}$$

This amplified and forward relay's end-to-end SNR is provided by:

$$\gamma_{AF} = \frac{P_t P_r \left|h_{SR}\right|^2 \left|h_{RD}\right|^2}{1 + P_t \left|h_{SR}\right|^2 + P_r \left|h_{RD}\right|^2} \tag{4.8}$$

4.3 Spatial Multiplexing

The channel must be shared by several users or applications in today's communication applications. Multiplexing is a technique for distributing multiple signals over a single channel. When several signals use the same source, they should be able to be distinguished from one another, i.e. orthogonal. Multiplexing comes in a number of ways. Here are a few examples:

- multiplexing of frequencies;
- multiplexing of wavelengths;
- multiplexing of time; and
- multiplexing in space.

We are specifically involved in space multiplexing which means to share a channel by focusing specific signals in non-overlapping narrow beams [85]. Spatial multiplexing splits the data source of a single individual into several sub-streams. Using an array of transmit antennas, all of the parallel sub-streams are transmitted continuously in the same frequency range. Since source data is transmitted in parallel through many antennas, the data rate improves.

V-BLAST (Vertical – Bell Labs Layered Space-Time) [86] is a Bell Labs spatial multiplexing architecture for obtaining exceptionally high throughput over fading wireless networks. It's used to make a contact between two points. If only one source can transmit the signal to the destination, the source splits the signal into various sub-streams and sends each sub-stream to the destination by multiple antennas. The receiver is equipped with multiple antennas to track all sub-streams. Each transmitting antenna receives a mixture of signals from each sub-stream. The sub-streams are scattered differently if multipath scattering is sufficient when they are transmitted from separate antennas at the transmitter site. The inevitable multipath in wireless networking has a very useful spatial parallelism that is used to greatly improve data rates. Multipath has been turned into a benefit. Advanced signal processing techniques such as successive interference cancellation (SIC) or maximum likelihood (ML) can be used to identify and recover the sub streams [86]. The technique relies on the receiver's BLAST signal processing algorithms. At the receiver, all of the signals from all of the receiver antennas are read at the same time. As the strongest sub stream has been removed, the remaining weaker signals have become easier to retrieve until the stronger signals are no longer a source of interference. The ability to differentiate between sub streams is dependent on distinctions in how they spread across the environment. It can be broken down into three easy steps: ordering, canceling, and nulling.

Since multiple antennas at the transmitter are not possible, a new method called cooperative spatial multiplexing has been introduced. The spatial multiplexing mechanism is aided by the relays. Cooperative spatial multiplexing, unlike traditional spatial multiplexing, does not require multiple antennas at the source. This scheme is shown in detail in Figure 4.2. In the thesis work, the AF scheme is used at the relays. From the relays to the destination, the system can be viewed as a V-BLAST scheme.

This approach is particularly helpful for uplink transmissions (from the mobile device to the base station). This scheme may be used in sensor networks, where even the source node could transfer the information symbol to the controller, while interacting with neighboring sensors to form a virtual array.

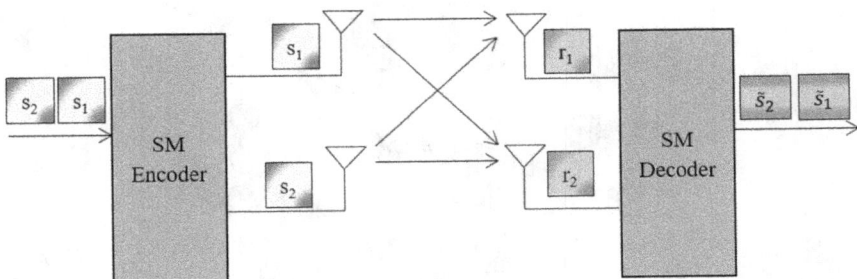

Figure 4.2 Spatial multiplexing.

4.4 System Model

The Multi-Radio Multi-Hop Cooperative Wireless Network (MRMHC) with N nodes, spread over an area of $LxLm^2$, as seen in Figure 4.3. Each node is fitted with a K radio terminal and a power control unit. Each node in the network employs amplify and forward (AF) relay protocol. The maximum transmitting power of the node is P_t. Let x be the data that is transmitted from node i to node j along two paths:

(i) Direct path $(i \rightarrow j)$; and

(ii) Cooperative path $(i \rightarrow l \rightarrow j)$.

The received information at node j via direct path y_{ij} and cooperative path y_{ij} are given by [87],

$$y_{ij} = \sqrt{P_t}\,h_{ij}x + \eta_{ij} \qquad (4.9)$$

$$y_{ij} = AP_t h_{ij}x + A\sqrt{P_t}\,h_{lj}\eta_{ij} + \eta_{ij} \qquad (4.10)$$

where h_{ij} is the channel coefficient of a link between i and j, η_{ij} is the AWGN between node i and j with zero mean and A is the

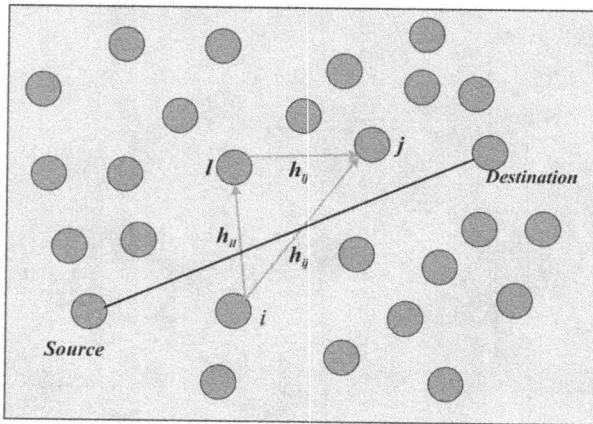

Figure 4.3 A MRMHC Network.

Amplification factor. Node j uses Maximum Ratio Combiner (MRC) technique to combine the information received from the two paths.

4.4.1 Cooperative Spatial Multiplexing

Motivated by the Mobile Communications Concept, we assume Line of Sight (LOS) to become the shortest path between these two nodes. We consider Rayleigh fading channel. In this section we also present a cooperative spatial multiplexing (CSM) scheme, to improve the network performance. Spatial diversity can be obtained in an MRMHC wireless network, since each node is equipped with multiple radios. The routing path between source and destination is calculated by using our routing algorithm proposed in Chapter 3. In that, each node transmits data using single radio terminal with fixed transmission power. In this chapter, spatial diversity is implemented by using multiple radio (K) terminals. Since the maximum transmit power of each node is P_t, the power control unit will distribute the power equally among K radio terminals, i.e., each radio terminal is allowed to transmit with the power $\dfrac{P_t}{K}$.

The amount of energy consumed per bit for i to j transmission (i.e., per hop E_b) is given by [17],

$$E_b = \sum_{i=1}^{K} (1+\alpha)\left(\frac{P_t}{K}\right) Q_0 \left(d_{ij}\right)^n M_l N_f + \frac{P_{Tx} + P_{Rx}}{b} \quad (4.11)$$

Where $Q_0 = \dfrac{(4\pi)^n}{G_{Tx} G_{Rx} \lambda^2}$; b is the bit rate, α $(=\frac{\zeta}{\xi}-1)$ is the transmission efficiency of a power amplifier which depends on the modulation. For MQAM $\zeta = 3\dfrac{2^{\frac{b}{2}-1}}{2^{\frac{b}{2}+1}}$, d_{ij} is the is the distance between the node i and j, G_{Tx} and G_{Rx} are the gain factors of transmitter and receiver respectively, n is the path loss exponent, M_l is the Link margin, N_f is the noise figure and, P_{Tx} and P_{Rx} are the transmitter and receiver circuit powers respectively.

The total energy consumption of a path from source to destination can be obtained by adding energy consumption of each hop.

$$E_{total} = \Sigma E_h \qquad (4.12)$$

4.4.2 Simulation Results

Simulation results of the proposed CSM scheme are presented in this sub-section. The parameters used for the simulation are listed in Table 4.1. The simulation is carried out using MATLAB®.

Figures 4.4 and 4.5 show that, with the same amount of power, we have achieved improved BER efficiency by applying spatial multiplexing in multi-radio networks. Here we have evaluated energy consumption of the network by varying number of hops and from the simulation results it can be observed that by implementing spatial multiplexing, better BER performance is achieved without increasing the power of the network.

Table 4.1 Simulation parameters for the proposed Cooperative Spatial Multiplexing scheme

PARAMETER	VALUE
CELL AREA L	1000 X 1000 M²
Target BER	10^{-3}
Transmitter and Receiver Gains	5dB
Transmitter Circuit Power P_{Tx}	97.8mW
Receiver Circuit Power P_{Rx}	119.8mW
Link Margin M_l	40dB
Noise Figure N_f	10dB
N_0	10^{-10}
Wavelength	0.12m
Number of Radio terminals K	2
Number of Nodes	500
Number of Flows	1

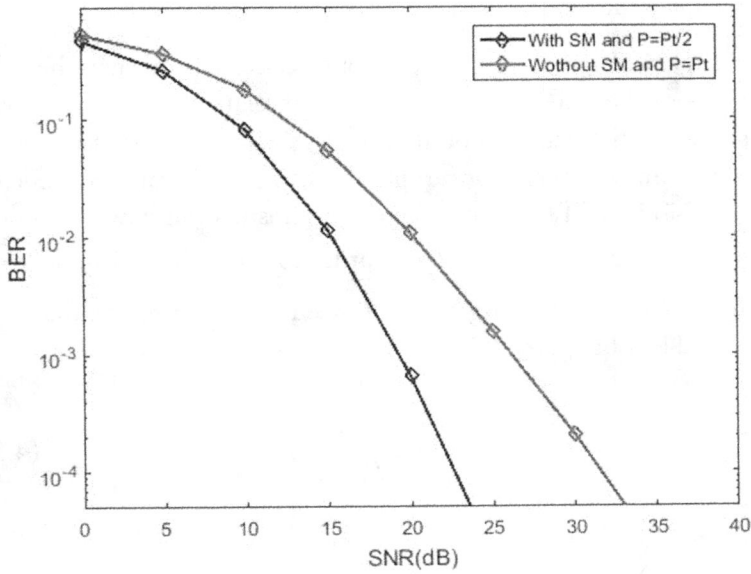

Figure 4.4 BER Performance with SM in multi-radio networks.

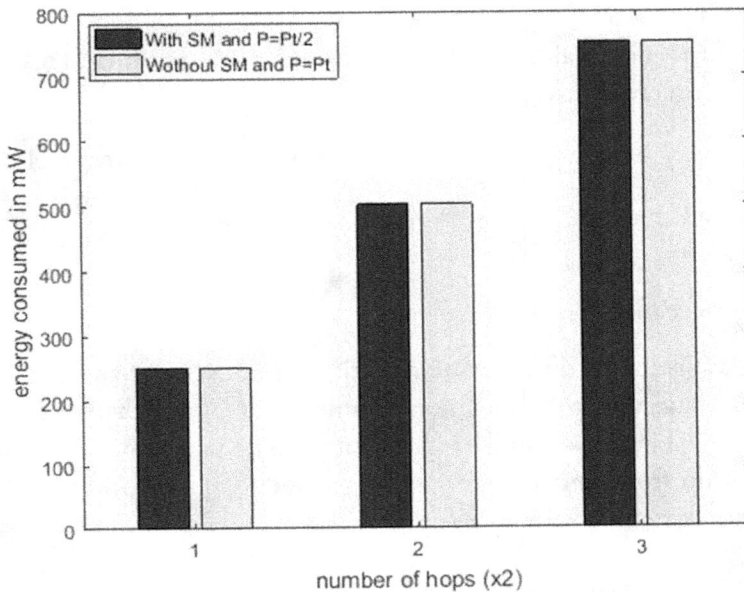

Figure 4.5 Energy comparison in mW by with SM in multi-radio networks.

4.5 Energy-Efficient Transmission

The simulation results in the previous section show that the BER efficiency of the CSM scheme is increased relative to the fixed power transfer with the same amount of energy consumption. In order to further minimize energy consumption, we suggest an Energy-Efficient Transmission (EET) scheme where the transmission power of a node can be minimized (less than P_t) in order to meet the BER P_b target.

The maximum transmitting power P needed to accomplish the target BER shall be obtained by the Chernoff bound.

$$P \leq \frac{KN_0}{P_b^{1/K}} \tag{4.13}$$

Where $P < P_t$

The EET scheme is described as follows:

- The routing path to the destination node is determined by the source node using the routing algorithm proposed in Chapter 3.
- Every node calculates P using Equation 4.13 in order to reach the target BER.
- Every node uses K radio terminals with a $\dfrac{P}{K}$ power level to transmit data.

4.5.1 Simulation Results

In this subsection, the simulation results of EET scheme are presented and we have used the simulation parameters of CSM scheme given in Table 4.1. Figures 4.6 and 4.7 represent the path taken by the proposed algorithm from source node 1 to target node 15, with 500 and 1000 nodes, respectively. The paths are as follows:

$$N = 500 : 1 \Rightarrow 23 \Rightarrow 177 \Rightarrow 188 \Rightarrow 349 \Rightarrow 343 \Rightarrow 15$$

$$N = 1000 : 1 \Rightarrow 545 \Rightarrow 910 \Rightarrow 115 \Rightarrow 873 \Rightarrow 15$$

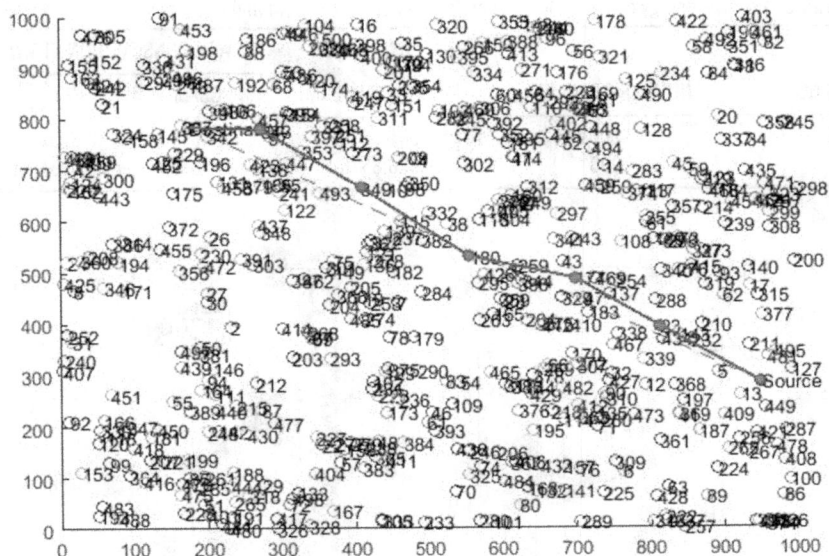

Figure 4.6 Path from node 1 to node 15 with 500 nodes.

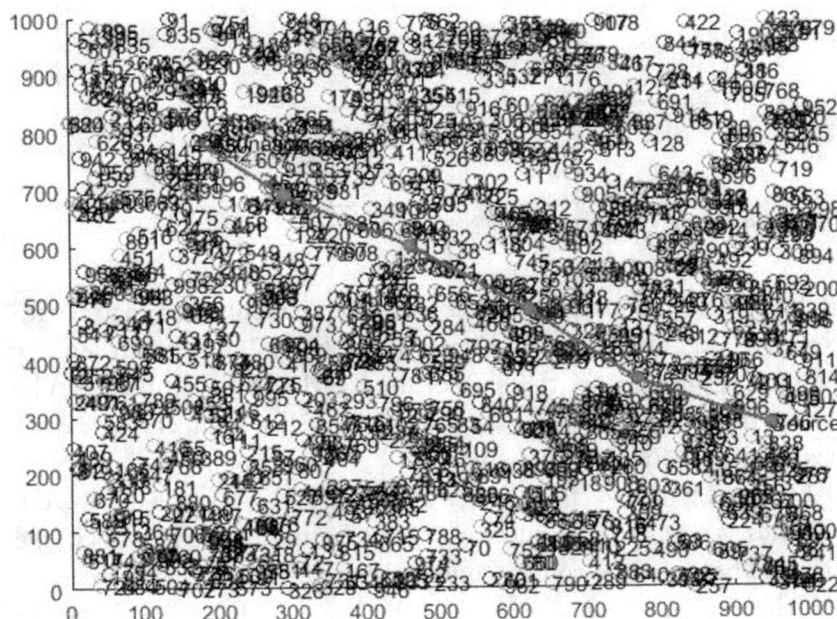

Figure 4.7 Path from node 1 to node 15 with 1000 nodes.

Table 4.2 The energy consumption values for the path of the CSM and EET

			POWER CONSUMPTION IN MW	
S.NO	NUMBER OF NODES	PATH DISTANCE (M)	CSAM	EET
1	500	945.6879	501.9659	0.6752
2	1000	926.8576	500.8573	0.5664

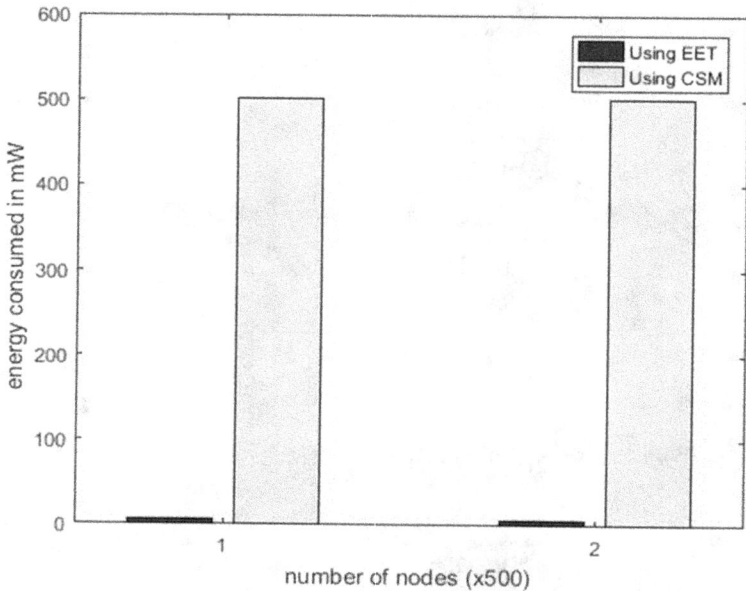

Figure 4.8 EET energy consumption with number of nodes.

The energy consumption of the above-mentioned path with N = 500 and 1000 is seen in Figure 4.8. Table 4.2 displays the energy consumption values for the path of the CSM and EET, and it can be shown that when following the proposed EET system, energy can be saved by up to 99 percent (approx.).

Figure 4.9 shows the energy consumption of CSM and EET schemes as a function of number of hops for normalized distance (the distance between the source node and the destination node is unity) and the energy consumption values are listed in Table 4.3. It can be observed that the energy consumed in EET scheme is approximately 91 percent of that of CSM scheme.

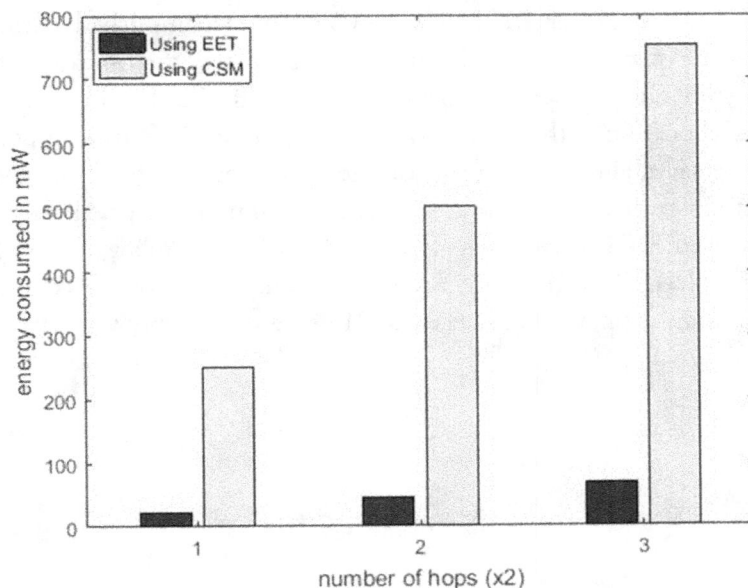

Figure 4.9 Energy consumption of CSM and EET schemes.

Table 4.3 Energy consumption values with number of hops

S.NO	NUMBER OF HOPS	POWER CONSUMPTION IN MW	
		CSM	EET
1	2	250.8742	21.76
2	4	501.7484	43.52
3	6	752.62	54.41

The proposed scheme saves about 99 percent of energy as opposed to fixed power transmission, according to the simulation results. The efficiency of the proposed EET is often evaluated in terms of the number of hops for a normalized distance between the source and destination. According to the simulation, we will save up to 91 percent of energy while we use the EET.

4.6 Summary

This chapter proposes an energy-efficient transmission scheme for large-scale multi-radio multi-hop cell ad hoc networks to reduce energy usage. Better BER efficiency can be accomplished without increasing network power by introducing spatial multiplexing. The suggested EET scheme utilizes spatial multiplexing and dynamic power allocation to the nodes to achieve the target BER. The proposed scheme saves about 99 percent of energy as opposed to fixed power transmission, according to the simulation results. The efficiency of the proposed EET is often evaluated in terms of the number of hops for a normalized distance between the source and destination. According to the simulation, we will save up to 91 percent of energy while we use the EET.

5

Energy-Efficient Routing Algorithm for Large-Scale Cooperative Mobile Ad Hoc Networks

5.1 Introduction

Cooperative communication (CC) is a useful feature for combating fading in wireless communication by using several radio terminals on the transmitter and/or receiver to create spatial diversity. Relay nodes retransmit the source's copy of the data, then pair it with a duplicate for improved original data decoding. With the advent of virtual MIMOs, CC technology enhances network efficiency in terms of output, reliability and power [89,90]. Researchers are interested in applying three-stage collaboration to large-scale networks and seeing how well it does. Increased connectivity between neighbors, on the other hand, contributes to increased congestion and, as a result, degrades network reliability much more than direct contact (without cooperative communication). With the advancement of modern wireless technologies, many radio terminals can be equipped with electronic devices that adhere to the IEEE Network Standard [91], lowering their costs. Interference may be reduced by allowing neighbor propagation through several orthogonal channels, which increases network bandwidth [92]. The use of resources is a big problem for MANETs. The need for large capacities is increasingly growing as a consequence of high data rate implementations, which would boost energy consumption, network lifetime and efficiency.

DOI: 10.1201/9781032714011-5

These problems are solved through clustering routing strategies [93]. The hierarchy of lower adaptive energy clusters (LEACH) is one of the most popular clustering techniques. In recent years, a number of improved LEACH routing schemes have been proposed, with a focus on changed cluster heads (CHs), network topology, and network extension [11]. The authors in [74] have proposed a virtual cooperative MIMO transmission system and derived an analytical term for the optimum number of cooperative nodes in two-stage cooperative networks. A low-complexity cooperative routing algorithm with an efficient power distribution approach was suggested in [12]. The authors of [13] suggested routing algorithms that optimize the efficiency of the physical, medium access control, and network layers to reduce network energy. The authors suggest a mutual automated repeat request (ARQ) process at the MAC layer to accomplish this. In [14], authors proposed a cooperative routing algorithm focused on quality of service to reduce energy usage. However, both of the above scholars took into account a network in which each node is fitted with a single radio terminal.

Muhammad Asshad et al. [17] use the max-min best relay selection method to analyze the efficiency of a non-regenerative wireless cooperative network over Rayleigh and Weibull fading networks. The Weibull fading parameter was used to extract the SNR moment generation function (MGF) at the destination node. The lower bound value of symbol error rate and outage likelihood was estimated using MGF.

Though [15] affects the accuracy of the analytical model due to a large number of fewer relay nodes, [16] major issues are routing and energy consumption, [18] the core network and wireless link can afford high pressure when the same data is applied by many users, [19] does not control the power of packet error rate (PER), maximizing the secrecy rate, [10] obtains a poor solution and low converge. As a consequence of the above problems, it is important to establish a new methodology for energy utilization routing algorithms in a cooperative network.

As a consequence, the key problem confronting large-scale MANET is energy consumption and routing, which must be overcome. Various methods have been applied, but owing to low efficiency, limited network lifetime, packet transmission issues, and high energy consumption, a remedy is yet to be found. As a result, a new approach for increasing network lifetime and lowering energy consumption is adopted.

5.2 Hybrid Multi-Hop Cooperative Routing Algorithm

In recent years, cooperative connectivity in wireless networks has increased in popularity as a way to alleviate the incredibly severe channel impairments induced by multipath propagation. To better refine the structure and enhance efficiency, MANET and cooperative communications are used. However, in MANETs energy consumption is a key problem. The need for high bandwidth is growing as a consequence of high data rate applications, which would boost energy consumption and shorten the network's lifetime. Mobile ad hoc networks are the purest type of autonomous structures, and as a result, cooperative networking faces various challenges. As a consequence, much ad hoc network analysis has centered on studying basic routing and clustering algorithms. Highly specialized protocols for embedded nodes have also been developed to minimize process energy consumption while still hitting the whole network with a large likelihood with the smallest period of time. As a result, there is a crucial need to create a new routing algorithm that uses fewer resources.

By integrating clustering and location-based routing techniques, a novel hybrid multi-hop cooperative routing algorithm for large-scale cooperative networks is proposed in this chapter. When a flow request is sent, cluster heads split the network into clusters. The Signal to Noise Ratio (SNR), relative distance, and relative mobility are all factors in the creation of a cluster. Following the formation of the cluster, one of the nodes is chosen as the cluster head depending on its position. To reduce end-to-end energy demand, maximize the number of transmitters and receivers in each hop and achieve an optimum number of cooperative relays in each hop. Figure 5.1 depicts the scenario.

5.2.1 Large-Scale Cooperative MANET

A Large Scale Cooperative Mobile Ad hoc Network (LC-MANET) is a network with N nodes uniformly spread over a region of

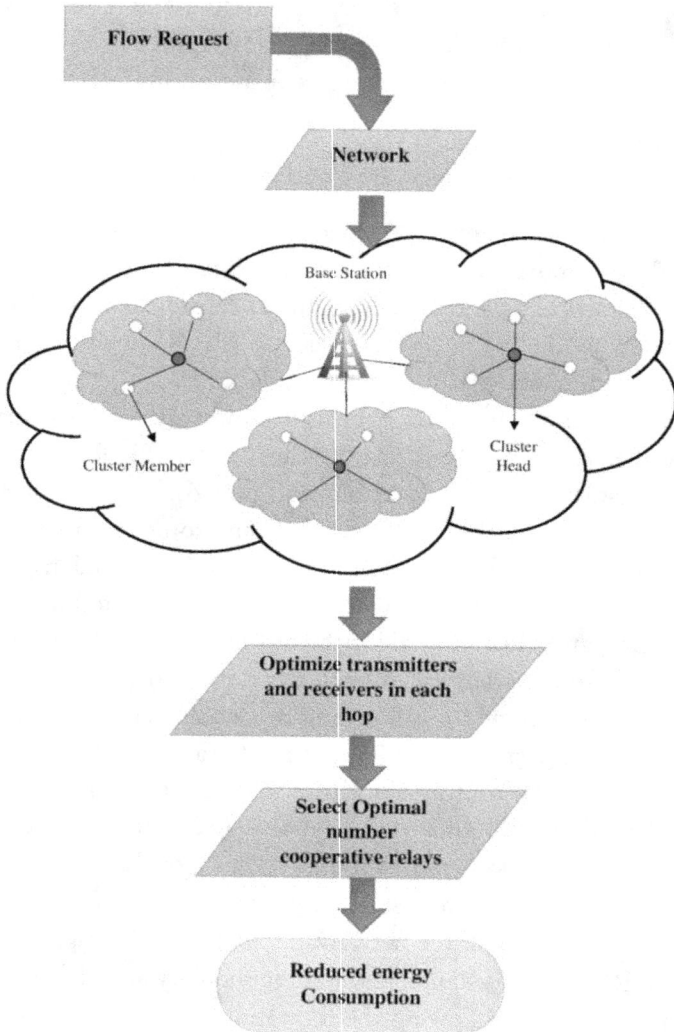

Figure 5.1 Hybrid multi-hop cooperative routing algorithm.

$L \times Lm^2$, as shown in Figure 5.2. The Decode and Forward (DAF) relay protocol is used by each node in the network, which is thought to be self-organized. Assume that each network node has K radio terminals and a power management system that varies transmitting power depending on size. R_i denotes the nodes in the transmitting

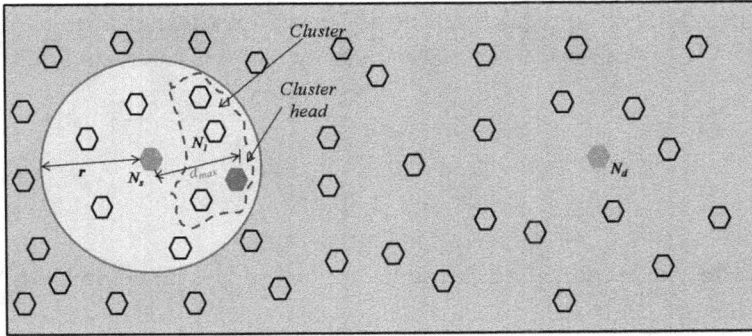

Figure 5.2 Large-scale cooperative MANET.

region of node i (N_i) that can interact directly with a likelihood of error $\left(P_e\right)$ less than or equal to a predefined threshold, while R and r denote the transmission coverage area and transmission radius, respectively.

Assume that almost all nodes throughout the LC-MANET are capable of encoding and decoding, complete channel estimation and synchronization; and maximum likelihood (ML) detection at the receiver. Please remember that the channel between the nodes is Rayleigh fading. Let's assume a node i broadcasts information X, which is successfully decrypted by another node $j \in R_i$. The obtained information y_i is provided by at node j.

$$y_i = \sqrt{P} h_{ij} X + \eta_j \qquad (5.1)$$

Where h_{ij} = is the channel coefficient between nodes i and j modeled as complex Gaussian random variable i.e., $\left|h_{ij}\right|^2 = \mu_{ij}^2 d_{ij}^{-4}$; μ_{ij}^2 and d_{ij} are the variance and distance between i and j; X is the compressed encoded data transmitted by node i and η_j represents zero-mean additive Gaussian noise with the variance σ^2.

Any node can get its location using GPS and the location of its neighbors by sharing beacon signals on a regular basis (every β sec). Any node obtains parameters such as link Signal to Noise Ratio

(SNR), distance, and relative velocity based on these beacon signals. The SNR of the relation between nodes p and q is calculated as:

$$\gamma_q = \frac{P\left|h_{pq}\right|^2}{\sigma_q^2} \tag{5.2}$$

Node p calculates the relative distance to node q based on the SNR as follows:

$$\Delta d_{pq} = \left(\frac{P\mu_{pq}}{\sigma_q \Delta\gamma_{pq}}\right)^{\frac{1}{4}} \tag{5.3}$$

Where $\Delta\gamma_{pq}$ is the nothing but the relative SNR, it is obtained as

$$\frac{1}{\Delta\gamma_{pq}} = \left|\frac{1}{\gamma_p^{t2}} - \frac{1}{\gamma_p^{t1}}\right| \tag{5.4}$$

and $t_2 - t_1 = \beta$. The relative velocity of the nodes is given as:

$$\Delta v_{pq} = \frac{\Delta d_{pq}}{\beta} \, m/sec \tag{5.5}$$

Each of the nodes in the cluster is chosen as a cluster head depending on its location after the cluster has been formed and the SNR has been calculated. The routing algorithm is explained in the following part.

5.3 Hybrid Cooperative Routing Mechanism for Low Energy Consumption

The proposed energy-efficient hybrid cooperative routing scheme for LC-MANET is defined first, followed by the optimization of cooperative nodes to reduce energy usage.

5.3.1 Cooperative Routing Algorithm

As a new flow enters from source node N_s to destination node N_d, node N_s searches for a group of nodes in its transmission coverage area and calculates the metrics; relative SNR and relative velocity, as defined in the system model, using periodically exchanged beacon signals. The source node forms a cluster and defines the Cluster head $\left(N_h\right)$ based on calculated metrics, where $N_h \in R_s$. The compressed encoded data $\tilde{X}$, as well as the destination and cluster head ID, are broadcast by the source node.

Algorithm: Energy-Efficient Hybrid Cooperative Routing

Input: A new flow arrival source (N_s) to destination N_d
Output: Routing path from source to destination with each next hop's Cluster Head and/or Cooperative relay nodes.
1: **While** source ≠ destination **do**
2: The source node measures the metrics using periodically exchanged beacon signals.
3: Find a set of nodes (R_s) in its transmission coverage area R.
4: **If** $N_d \in R_s$ **then**
5: Cluster head=destination
6: **Else**
7: The source forms a cluster with the nodes which are having the relative velocity (with source) less than a predefined threshold i.e.,

$$V_h = \left\{ l \,\middle|\, \left(\max\left(\Delta v_{si}\right) - \Delta v_{si}\right) < v_r \,\middle|\, l \in \left(R_h - R_{h-1}\right)\bigcup N_{h-1} \right\}$$

8: From the above cluster, the source selects cluster head for the next hop as:

$$N_h = \underset{l \in V_h}{argmax}\left\{ d_{N_{h-1},l} \right\} \forall h \geq 2$$

9: **end if**
10: The other nodes in the cluster cooperate with the source to forward the data to the cluster head.
11: Source node= cluster head
12: **end while**

It denotes h^{th} hop cluster head and set of cluster nodes as N_h and V_h respectively. The $(h+1)^{th}$ transmission is required only when the destination node is not in the transmission range of N_h i.e., $N_d \notin R_{N_h}$.

5.3.2 Energy Consumption Analysis

A cooperative MISO transmitting scheme and an energy consumption model for a single hop is presented in this segment. The optimum number of cooperative nodes was determined using this model. The data is transmitted in two phases by the source node (N_s), which forms a cluster.

- *Phase I:* It broadcasts the data to all nodes in the cluster in the first step. Assume that the cluster contains n nodes. MQAM modulation's average energy usage can be represented as [22]:

$$E_{P_1} = \frac{\xi}{\eta} Q_0 E_{b,P_1} r^2 + \left(P_{tx} + nP_{rx} \right) \Big/ bB \qquad (5.6)$$

Where $Q_0 = \dfrac{(4\Pi)^2 M_l N_f}{G_{tx} G_{rx} \lambda^2}, \xi = 3\dfrac{2^{b/2} - 1}{2^{b/2} + 1}; G_{tx} \, and \, G_{rx}$ are gains of receiver and transmitter respectively.

M_l is the link margin, N_f is the receiver noise figure, λ is the carrier wavelength, $\overline{E_{b,P_1}}$ the average received energy pet bit in phase 1, b is the transmission bit rate, B is the modulation Bandwidth, $P_{tx} \, and \, P_{rx}$ are the transmitter and receiver circuit powers respectively.

The average number of nodes in the cluster is

$$n = \frac{\pi r^2 N}{L^2} P(\Delta v) \qquad (5.7)$$

After phase 1 has broadcasted the cluster, phase 2 is processed with n nodes in order to transmit data to the cluster head.

- *Phase II:* The data is distributed to the next-hop cluster head by n nodes (n-1 cluster nodes and source node) in step 2. The average energy intake can be calculated as follows:

$$E_{P_2} = \frac{\xi}{\eta} Q_0 \overline{E_{b,P_2}} d_{max}^2 + \left(nP_{tx} + P_{rx} \right) \Big/ bB \qquad (5.8)$$

The average energy consumption per bit of every hop is $E_b = E_{P_1} + E_{P_2}$

Chertoff upper bound with several transmitting antennas equal to one may be used to achieve the upper bound $\overline{E_{b,P_2}}$.

$$\overline{E_{b,P_2}} \leq \frac{2(2^b - 1)N_0 n}{3b} \left(\frac{4}{bP_e}\right)^{1/n} \tag{5.9}$$

$\overline{E_{b,P_2}}$ can be obtained by substituting n=1. Using equation 5.9 as an approximation, the following closed-form expression for the average energy consumption per bit was obtained:

$$E_b = C_b \left[\frac{C_e L^2}{\pi N P(\delta v)} + (C_e) \frac{1}{n} d_{max}^2 \right] n + C_p (n+1) \tag{5.10}$$

Where, $C_b = \dfrac{\xi Q_0 2(2^b - 1) N_0}{3b\eta}, C_e = \dfrac{4}{bP_e}$ and $C_p = \dfrac{P_{tx} + P_{rx}}{bB}$. As per the proposed algorithm, the bth Hop cluster head should be within the hop cluster head's transmission coverage range $(b-1)^{th}$. Therefore the distance between two cluster heads should be (d_{max}) should be $d_{max} \leq r$. The cluster's average number of nodes becomes.

$$n \leq \frac{\pi N d_{max}^2 P(\delta v)}{L^2} \tag{5.11}$$

Where, $P(\Delta v)$ the probability of the node having relative mobility difference is less than the threshold. Calculate a close approximation of the ideal value of n to reduce average energy consumption per bit $E(b)$ when $d_{max}^2 \geq \dfrac{nL^2}{\pi N P(\Delta v)}$ as:

$$\min_{n} E_{b}, \, s.t. 2 \leq \frac{\pi N d_{\max}^{2} P(\Delta v)}{L^{2}} \tag{5.12}$$

Otherwise, in the SISO transmission scheme, n=1 sends the data. Differentiating with respect to n yields the essential value of a function E_{b}:

$$d_{\max}^{2}\left(C_{e}\right)^{1/n}\left[n - 1n\left(C_{e}\right)\right] + \left[\frac{C_{e}L^{2}}{\pi N P(\delta v)} + \frac{C_{p}}{C_{b}}\right] = 0 \tag{5.13}$$

Since the equation above this is positive, n must be less than $\ln\left(C_{e}\right)$. Let the above equation's positive real-valued solution be n_{p}. The optimal value E_{b} is then calculated as follows:

$$n_{0} = \begin{cases} \lfloor n_{p} \rfloor & if \, 2 \leq n_{p} \leq \dfrac{\pi N d_{\max}^{2} P(\Delta v)}{L^{2}} \\ 2 & if \quad n_{p} < 2 \end{cases} \tag{5.14}$$

For the study of multi-hop networks, the suggested routing algorithm reduces energy consumption.

5.4 Results and Discussion

This section compares and contrasts the experimental findings obtained with conventional approaches to demonstrate the viability of the proposed process. The tools for specification and implementation are described below.

The proposed methodology is outlined in depth in Section 5.3, and its output is evaluated in this section. With the following framework definition, the suggested approach is introduced in the MATLAB working platform. This section presents a simulation analysis of the proposed algorithm. Simulate the algorithm in MATLAB with the following parameters as in Table 5.1.

Table 5.1 Parameters for simulation

NOTATION	MEANING	VALUE
N	NUMBER OF NODES	[100 1000]
P	Transmitted power	1Mw
N_0	Noise power spectral density	-171dBm/Hz
B	Modulation bandwidth	10KHz
	Combining strategy	MRC
β	Periodic interval	1μs
M_l	Link margin	40Db
N_f	Noise figure	10Db
P_e	Target BER	10^{-3}
G_{tx}, G_{rx}	Transmitter and receiver gain	5dBi
P_{tx}	Transmitter circuit power consumption	97.8Mw
V_r	Velocity threshold	5m/sec
P_{rx}	Receiver circuit power consumption	119.8Mw

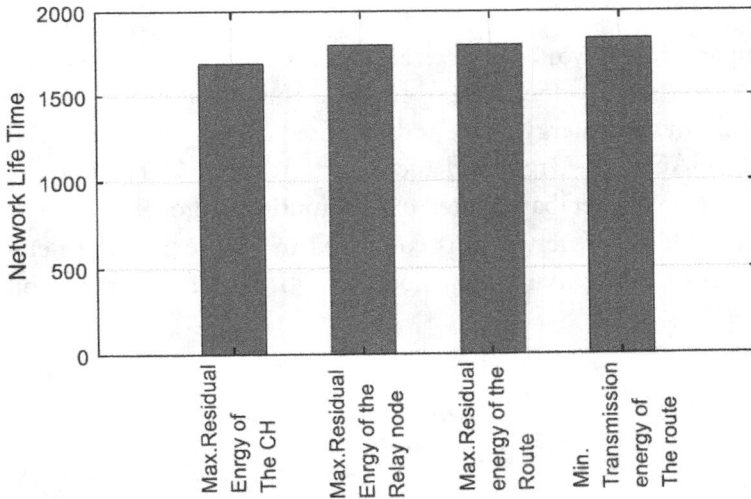

Figure 5.3 The fitness function's four parameters are compared in terms of network lifetimes.

Maximum residual energy of the cluster head, maximum residual energy of the relay node, maximum residual energy of the path, and minimum transmission energy of the route are shown in Figure 5.3. Criterion 4 has the longest lifetime. This means that criterion 4 has a load-balancing fairness test. As a result, the fitness function for hybrid multi-hop cooperative routing should be criterion 4.

Table 5.2 Network lifetime

FITNESS FUNCTION	NETWORK LIFE TIME
MAX. RESIDUAL ENERGY OF THE CH	1690
Max. Residual energy of the relay node	1800
Max. Residual energy of the route	1800
Min. Transmission energy of the route	1840

The network lifetime of the fitness function is seen in Table 5.2, where the maximum residual energy of the CH is 1690. The network lifetime residual energy of the relay node is 1800, Max. The route's residual energy over the network's lifespan is 1800, with a minimum of 1800. The network lifetime of the route's transmitting energy is 1840.

5.4.1 Comparison Results

Compare the network lifetimes and residual energies of sensor nodes in EECC (energy-efficient cooperative communication method) [94], HEED (hybrid-energy efficient distributed clustering approach) [95], and SOSAC (Self-Organized and Smart-Adaptive Clustering) [20], a well-known cluster-based inter-cluster routing protocol.

The lifetime of networks is compared in Figure 5.4. The network lifetimes of EECC (with relay nodes), SOSAC (without relay nodes),

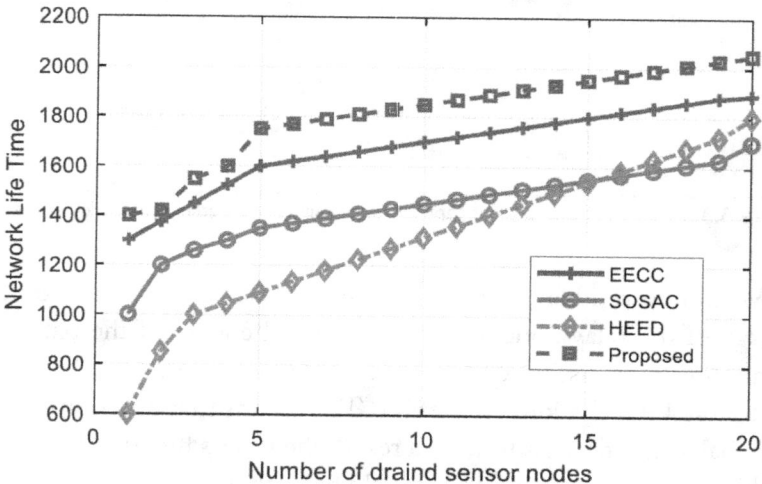

Figure 5.4 Comparison of the proposed algorithm's network lifetime.

Table 5.3 Comparison of network lifetime

NUMBER OF DRAINED NODES	NETWORK LIFETIME			
	EECC	SOSAC	HEED	PROPOSED
1	1300	1000	600	1400
2	1375	1200	850	1420
3	1450	1260	1000	1550
4	1525	1300	1045	1600
5	1600	1350	1090	1750
6	1620	1370	1135	1770
7	1640	1390	1180	1790
8	1660	1410	1225	1810
9	1680	1430	1270	1830
10	1700	1450	1315	1850
11	1720	1470	1360	1870
12	1740	1490	1405	1890
13	1760	1510	1450	1910
14	1780	1530	1495	1930
15	1800	1550	1540	1950
16	1820	1570	1585	1970
17	1840	1590	1630	1990
18	1860	1610	1675	2010
19	1880	1630	1720	2030
20	1890	1700	1800	2050

and HEED are compared to show the validity of the cooperative communication method.

Table 5.3 shows that the network lifetime of EECC is between 1300 and 1890 until the first node is depleted, which is far longer than the network lifetime of SOSAC. SOSAC has a lifetime of 1000 when the first node is drained and 1700 when the last node is drained. The first node drained value for HEED is 600, while the last node drained value is 1800. However, as opposed to the aforementioned methods, the proposed work drains the first node's lifetime to 1400 and the last node's lifetime to 2050. This experiment shows that CHs and relay nodes cooperating to reduce energy consumption and sustain load balancing improves the network lifetime.

Figure 5.5 displays a graph comparing the residual energy ratios of different approaches with the proposed algorithm, namely EECC (using relay nodes) and HEED.

The average ratios of the residual energies of EECC, HEED, and the suggested method are shown in Table 5.4. When the first node is

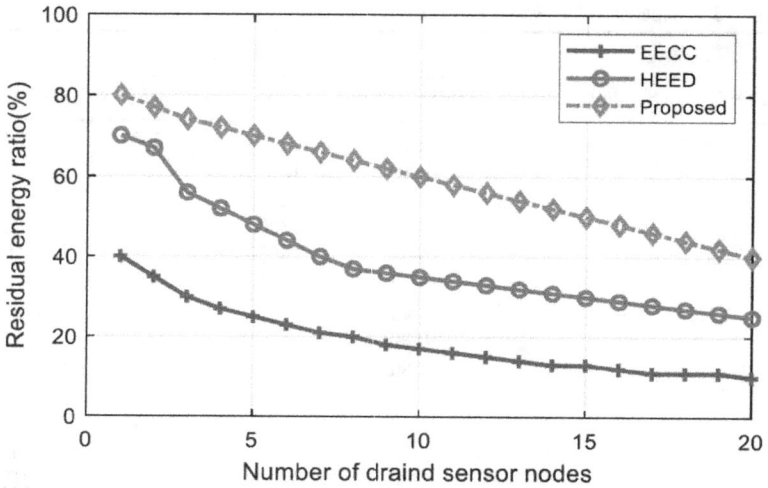

Figure 5.5 Experiment with the residual energy ratios in different methods.

Table 5.4 Comparison of residual energy ratio

NUMBER OF DRAINED SENSOR NODES	RESIDUAL ENERGY RATIO (%)		
	EECC	HEED	PROPOSED
1	40	70	80
2	35	67	77
3	30	56	74
4	27	52	72
5	25	48	70
6	23	44	68
7	21	40	66
8	20	37	64
9	18	36	62
10	17	35	60
11	16	34	58
12	15	33	56
13	14	32	54
14	13	31	52
15	13	30	50
16	12	29	48
17	11	28	46
18	11	27	44
19	11	26	42
20	10	25	40

drained, the total residual energy ratio of EECC is between 40 percent and 10 percent (when the 20th node is drained). And there's HEED, which has a ratio of 70 percent (when the first node is drained) to 25 percent (when the 20th node is drained). As compared to the existing system, the proposed method saves more resources, with a ratio of 80 percent (when the first node is drained) to 40 percent (when the 20th node is drained).

The proposed approach was also compared to existing methods such as EECC, HEED, and SOSAC, which are well-known for a cluster-based inter-cluster routing protocol that deals with network lifetime and residual energies of sensor nodes. As a consequence, certain techniques will be used to equate to the proposed method.

Figure 5.6 depicts network throughput, with Figures 5.6(a) depicting throughput versus pause time, 5.6(b) depicting throughput versus certain nodes, 5.6(c) depicting throughput versus CBR connection, and 5.6(d) depicting throughput versus packet size. The

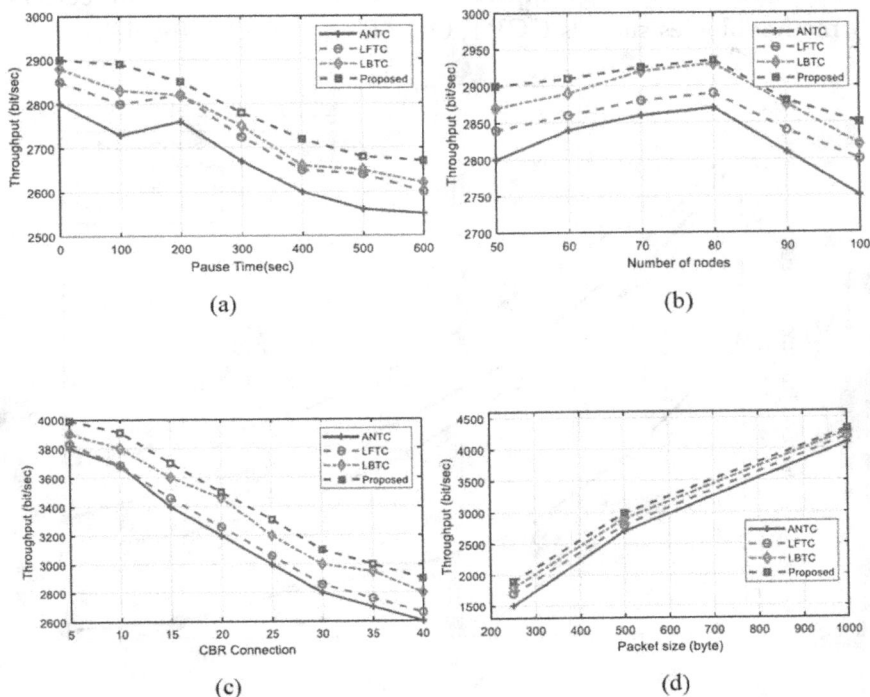

Figure 5.6 Network throughput over (a) Pause time; (b) Number of nodes; (c) CBR connection; and (d) Packet size.

strategies under consideration are ANTC (Adaptive Neighbor-based Topology Control), LFTC (Learning-based Fuzzy-logic Topology Control), and LBTC (Learning-based Fuzzy-logic Topology Control) (Location-Based Topology Control with Sleep Scheduling). According to Figure 5.7, the proposed scheme will achieve higher throughput. Higher throughput is related to the network's increased longevity as a result of adequate transmitting power modification.

The following methods are used to compare the proposed system to an existing system for multiple hops and energy consumption: ad hoc on-demand distance vector (AODV) routing algorithm [21], Opportunistic Cooperative Packet Transmission (OCPT) [22].

Figure 5.7 depicts the End-to-End Delay versus Node Count for methods such as ANTC, LFTC, and LBTC. As per Figure 5.7, LBTC has a greater end-to-end delay than LFTC and ANTC. This is attributed to a rise in hop count as nodes transmit at lower power levels.

In terms of End-to-end Delay versus Multiple Nodes, Figure 5.7 and Table 5.5 compare the proposed approach to current methodologies such as CCPT, OCPT, and AODV. The delay value

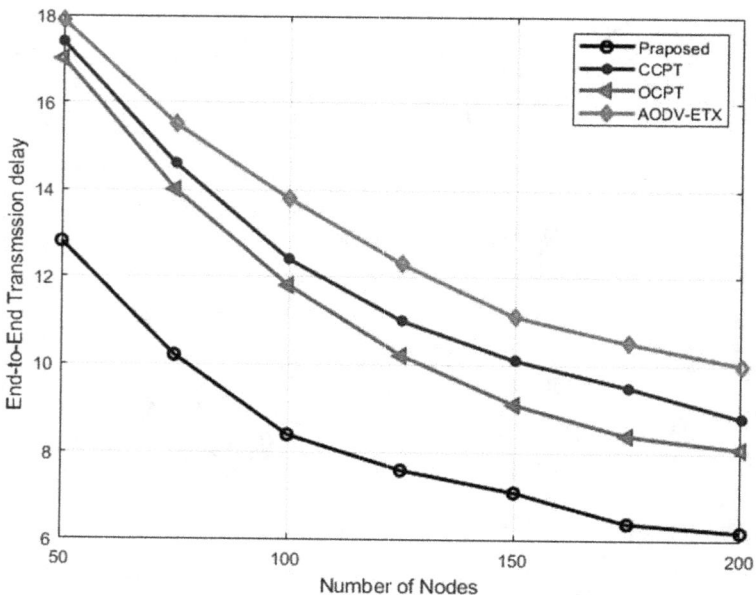

Figure 5.7 Delay Vs number of nodes.

Table 5.5 End-to-end delay comparison

NO. OF NODES	END-TO-END TRANSMISSION DELAY (MS)			
	PROPOSED	OCPT	CCPT	AODV
50	12.8	17	17.4	17.9
100	8.4	11.8	12.4	13.8
150	7.2	9.1	10.1	11.1
200	6.2	8.1	8.8	10

varies depending on the node, and the current approach has a lower delay value than the existing methods. The time difference for OCPT is 0.017 seconds for the 50th node and 0.0.0118 seconds for the 100th node. The delay time for CCPT is 17.4m seconds for the 50th node and 12.4m seconds for the 100th node. The delay time for AODV is 17.9m seconds for the 50th node and 13.8m seconds for the 100th node. At the 50th node, the proposed method took just 12.8m seconds, and at the 100th node, it took just 8.4m seconds. Furthermore, the existing method is a perfect fit for the comparison, yielding a nearly better outcome for throughput and delay to the proposed method.

Figure 5.8 depicts the total number of hops for different routing systems as a function of node count. It can be shown that the proposed routing scheme needs fewer hops than the AODV and OCPT systems, since as node density rises in the network, the chance of having a node away from the source increases, allowing the information to be forwarded to the destination in the minimal path length, i.e., in the fewest number of hops.

Figure 5.9 illustrates a study of end-to-end energy usage with different routing systems depending on the number of nodes. Since the work obtains an optimum number of cooperative nodes in each hop, the path's energy consumption would decrease. With increased node density, the proposed algorithm necessitates less hops and consumes less energy. At N=700 and L=1000, as compared to conventional AODV routing algorithms, proposed method will save 53.42 percent of the energy. The results in both existing methods are virtually identical, and the hops and energy consumption are performed in this process, which is related to the proposed method for evaluation.

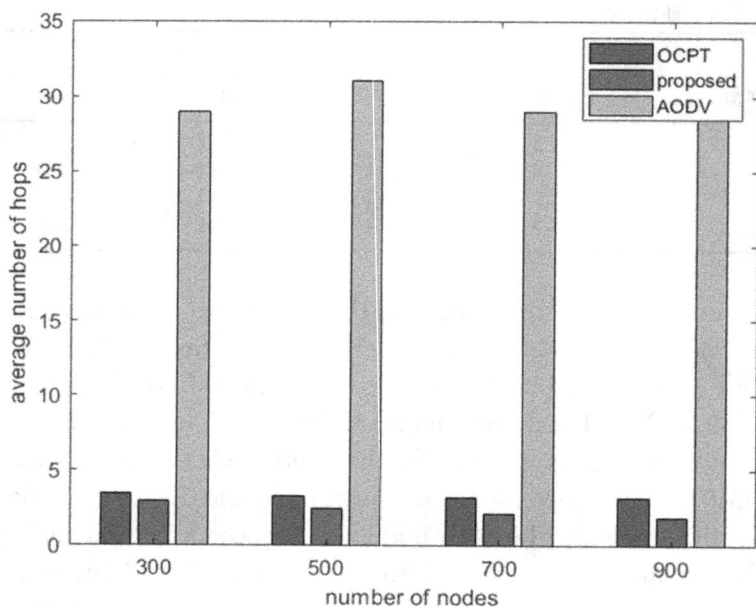

Figure 5.8 Average number of hops.

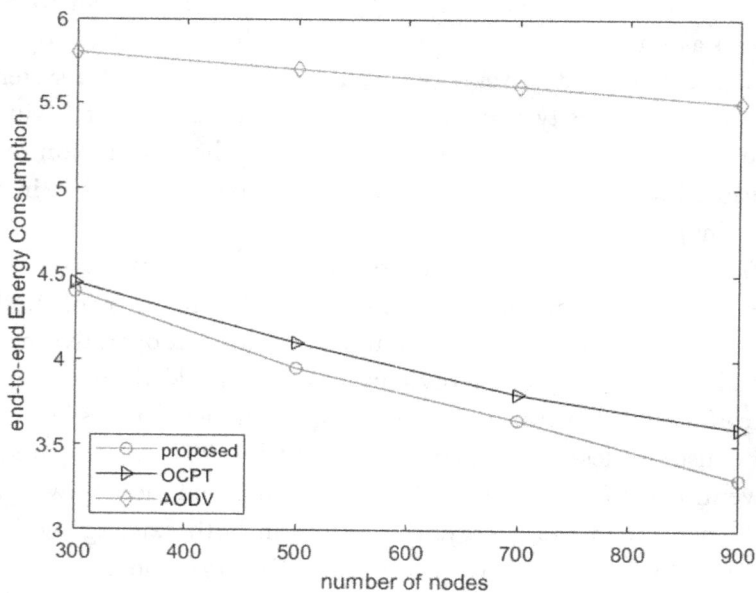

Figure 5.9 End-to-end energy consumption.

The proposed approach outperforms acceptable prior approaches in terms of network lifetime, throughput, delay, and energy usage. As opposed to the EECC, SOSAC, and HEED strategies, the network lifetime of the proposed work drained is larger, with the first node lifetime of 1400 and the last node lifetime of 2050. Furthermore, the proposed method ratio's residual energy is between 80 percent (when the first node is drained) and 40 percent (when the 20th node is drained), saving more energy than EECC and HEED strategies. In comparison to ANTC, LFTC, and LBTC, the suggested scheme achieves higher throughput and lower delay period. Higher throughput is related to the network's increased longevity as a result of adequate transmitting power modification.

Furthermore, the proposed routing system needs less hops than the AODV and OCPT systems because as the node density in the network increases, the probability of getting a node away from the source grows as well, enabling the information to be routed to the destination in a reduced path length, i.e., a minimum number of hops, while also reducing the energy consumption of the path. With improved node density, the proposed algorithm needs less hops, and the energy usage is also smaller. When compared to traditional AODV and OCPT routing algorithms, resources can be saved by 53.42 percent. As a consequence, the proposed approach increases the network lifetime thus lowering energy consumption.

5.5 Summary

Cooperative communications enable communication network nodes or terminals to collaborate in the transmission of information, making for more effective usage of communication resources. A hybrid multi-hop cooperative routing algorithm for LC-MANET was described in this work. To counteract the mobility impact and reduce the total number of hops, combine clustering and location-based techniques. Incorporated optimization mechanisms in each hop and obtained the optimum number of cooperative nodes by maximizing the number of transmitters and receivers simultaneously. When opposed to the conventional routing approach, simulation results indicate that the proposed algorithm saves up to 53.42 percent in energy consumption.

References

[1] T. Rappaport, (2001) *Wireless Communications: Principles and Practice.* Upper Saddle River, Second Edition, Pearson India Education Services Pvt. Ltd.

[2] Z. Lizhong and D. N. C. Tse, (2003) Diversity and Multiplexing: A Fundamental Tradeoff in Multiple-Antenna Channels, *IEEE Transactions on Information Theory*, vol. 49, No. 5, 1073–1096.

[3] J. N. Laneman, D. N. C. Tse, and G. W. Wornell, (2004) Cooperative Diversity in Wireless Networks: Efficient Protocols and Outage Behavior, *IEEE Transactions on Information Theory*, vol. 50, No. 12, 3062–3080.

[4] M. Dong, M. Hajiaghayi, and B. Liang, (2012) Optimal Fixed Gain Linear Processing for Amplify-and-Forward Multichannel Relaying, *IEEE Transactions on Signal Processing*, vol. 60, No. 11, 6108–6114.

[5] J. Guerreiro, R. Dinis, and P. Montezuma, (2014) Analytical Evaluation of Nonlinear Amplifyand-Forward Relay Systems for OFDM Signals. In: *IEEE 80th Vehicular Technology Conference*, pp. 1–5, Sep 2014.

[6] D. Kim, J. Seo, and Y. Sung, (2013) Filter-and-Forward Transparent Relay Design for OFDM Systems, *IEEE Transactions on Vehicular Technology*, vol. 62, No. 9, 4392–4407.

[7] T. M. Cover and A. A. El Gamal, (1979) Capacity Theorems for the Relay Channel, *IEEE Transactions on Information Theory*, vol. 25, No. 5, 572–584.

[8] G. Kramer, M. Gastpar, and P. Gupta, (2005) Cooperative Strategies and Capacity Theorems for Relay Networks, *IEEE Transactions on Information Theory*, vol. 51, No. 9, 3037–3063.

[9] A. Raja and P. Viswanath, (2014) Compress-and-Forward Scheme for Relay Networks: Backword Decoding and Connection to Bisubmodular Flows, *IEEE Transactions on Information Theory*, vol. 60, No. 9, 5627–5638.

[10] D. Gupta, A. Khanna, L. SK, K. Shankar, V. Furtado, and J. J. Rodrigues, (2019) Efficient Artificial Fish Swarm-based Clustering Approach on Mobility Aware Energy Efficient for MANET, *Transactions on Emerging Telecommunications Technologies*, vol. 30, No. 9, 1–10.

[11] N. Sabor and M. Abo-Zahhad, (2020) A Comprehensive Survey of Intelligent-Based Hierarchical Routing Protocols for Wireless Sensor Networks. *Nature-Inspired Computing for Wireless Sensor Networks* (pp. 197–257). Springer, Singapore.

[12] R. Varatharajan, A. P. Preethi, G. Manogaran, P. M. Kumar, and R. Sundarasekar, (2018) Stealthy Attack Detection in Multi-channel Multi-radio Wireless Networks, *Multimedia tools and applications*, vol. 77, No. 14, 18503–18526.

[13] S. Li, Z. Dou, F. Wang, and Q. Xu, (2019) The Energy Efficiency of Five broadcast-based ARQ Protocols in Multi-hop Wireless Sensor Networks, *IET Communications*, vol. 13, No. 15, 2243–2253.

[14] L. L. Hung, F. Y. Leu, K. L. Tsai, and C. Y. Ko, (2020) Energy-Efficient Cooperative Routing Scheme for Heterogeneous Wireless Sensor Networks, *IEEE Access*, vol. 8, No. 1, 56321–56332.

[15] Y. Gao, G. Kang, and J. Cheng, (2019) An Opportunistic Cooperative Packet Transmission Scheme in Wireless Multi-Hop Networks. In: *IEEE Wireless Communications and Networking Conference Workshop (WCNCW) 2019:*15–18.

[16] J. Cheng, Y. Gao, N. Zhang, and H. Yang, (2019) An Energy-efficient Two-stage Cooperative Routing Scheme in Wireless Multi-hop Networks, *Sensors,* vol. 19, No. 5, 1–15.

[17] M. Asshad, A. Kavak, K. Küçük, and S. A. Khan, (2020) Using Moment Generating Function for Performance Analysis in Non-Regenerative Cooperative Relay Networks with Max-Min Relay Selection. *AEU–International Journal of Electronics and Communications*, vol. 116, 153066.

[18] L. Boje, R. Wang, Y. Cui, Y. Gong, and H. Tan, (2020) Joint Optimization of File Placement and Delivery in Cache-Assisted Wireless Networks with Limited Lifetime and Cache Space. *arXiv preprint* arXiv:2001.02557 2020.

[19] Y. Gao, H. Ao, W. Zhou, S. Hu, H. Yu, Y. Guo, and J. Cao, (2019) A Novel AI-Based Optimization of Node Selection and Information Fusion in Cooperative Wireless Networks. In *Proceedings of SAI Intelligent Systems Conference 2019:*14–23.

[20] M. Elhawary, and Z.J. Haas, (2011) Energy-efficient Protocol for Cooperative Networks, *IEEE/ACM transactions on networking*, vol. 19, No. 2, 561–574.

[21] F. Ullah, Z. Ullah, S. Ahmad, I. U. Islam, S. U. Rehman, and J. Iqbal, (2019) Traffic Priority-based Delay-aware and Energy-efficient Path Allocation Routing Protocol for Wireless Body Area Network. *Journal of Ambient Intelligence and Humanized Computing*, vol. 10, No. 10, 3775–3794.

[22] J. Wang, Y. Gao, K. Wang, A. K. Sangaiah, and S. J. Lim, (2019) An Affinity Propagation-based Self-adaptive Clustering Method for Wireless Sensor Networks, *Sensors*, vol. 19, No. 11, p.2579.

[23] M. Singh and T. Singh, (2017) Energy-Efficient, Distributed Clustering Approach for Ad Hoc Wireless Sensor Network, *International Journal of Science and Research*, vol. 6, No. 4, 2415–2421.

[24] R. Tariq, S. Ahmed, R. S. Sani, Z. Najam, and S. Shafique, (2019) Securing Ad hoc On-demand Distance Vector Routing Protocol against the Black Hole DoS Attack in MANETs. *PeerJ PrePrints*, 7, p.e27905v1.

[25] Y. Gao, G. Kang, and J. Cheng, (2019) An Opportunistic Cooperative Packet Transmission Scheme in Wireless Multi-Hop Networks, *Sensors*, vol. 19, No. 21, 4821.

[26] J. Yindi and H. Jafarkhani, (2009) Single and Multiple Relay Selection Schemes and their Achievable Diversity Orders, *IEEE Transactions on Wireless Communications*, vol. 8, No. 3, 1414–1423.

[27] Z. Y. Liu, (2013) Single and Multiple Relay Selection for Cooperative Communication under Frequency Selective Channels. In: *IEEE Region 10 Conference, pp. 1–4, Oct 2013*.

[28] N. E. Wu, W. C. Huang, and H. J. Li, (2009) A Novel Relay Selection Algorithm for Relaying Networks. In: *IEEE 70th Vehicular Technology Conference, pp. 1–5, Sep 2009*.

[29] C. Yu, O. Tirkkonen, and J. Hamalainen, (2010) Opportunistic Relay Selection with Cooperative Macro Diversity, *EURASIP Journal on Wireless Communications and Networking*, vol. 2010, No. 1, pp. 1–14.

[30] J. N. Laneman and G. W. Wornell, (2003) Distributed Space-Time-Coded Protocols for Exploiting Cooperative Diversity in Wireless Networks, *IEEE Transactions on Information Theory*, vol. 49, No. 10, 2415–2425.

[31] J. Yang, Z. Zhang, and W. Meng, (2010) A Novel Relay Selection Scheme in Multi-Antenna Cooperative Systems. In: *IEEE International Conference on Software Engineering and Service Sciences, pp. 427–430, Jul 2010*.

[32] Q. Deng and A. G. Klein, (2012) Relay Selection in Amplify-and-Forward Relay Networks with Frequency Selective Fading. In: *46th Asilomar Conference on Signals, Systems and Computers, pp. 1356–1360, Nov 2012*.

[33] Q. Zhao and H. Li, (2005) Performance of Differential Modulation with Wireless Relays in Rayleigh Fading Channels, *IEEE Communications Letters*, vol. 9, No. 4, 343–345.

[34] A. Keshavarz-Haddad and M. A. Khojastepour, (2009) On Capacity of Deterministic Wireless Networks under Node Half-duplexity Constraint. In: *47th Annual Allerton Conference on Communication, Control, and Computing (Allerton)*, pp. 887–891, Sep. 2009.

[35] Y. G. Kim and N. C. Beaulieu, (2014) Relay Advantage Criterion for Multi-hop Decode and Forward Relaying Systems, *IEEE Transactions on Wireless Communications*, vol. 13, No. 4, 1988–1999.

[36] M. El-Aaasser and M. Ashour, (2013) Energy Aware Classification for Wireless Sensor Networks Routing. In: *15th International Conference on Advanced Communication Technology (ICACT)*, pp.66, 71, 2013.

[37] A.G. Marques, W. Xin, and G.B. Giannakis, (2008) Minimizing Transmit Power for Coherent Communications in Wireless Sensor Networks With Finite-Rate Feedback, *IEEE Transactions on Signal Processing*, vol. 56, No. 9, 4446–4457.

[38] C. Long, H. Chen, and L. Li, (2008) Energy-Efficiency Cooperative Communications with Node Selection for Wireless Sensor Networks. In: *Pacific-Asia Workshop on Computational Intelligence and Industrial Application, (PACIIA '08)*, pp. 761, 765, 2008.

[39] L. Viet-Anh, R. A. Pitaval, S. Blostein, T. Riihonen, and R. Wichman, (2010) Green Cooperative Communication Using Threshold-based Relay Selection Protocols. In: *International Conference on Green Circuits and Systems (ICGCS)*, pp. 521–526, 2010.

[40] C. Dan, J. Hong, and L. Xi, (2011) An Energy-Efficient Distributed Relay Selection and Power Allocation Optimization Scheme over Wireless Cooperative Networks. In: *IEEE International Conference on Communications (ICC)*, pp. 1–5, 2011.

[41] H. Kim, S.-R. Lee, C. Song, K.-J. Lee, and I. Lee, (2015) Optimal Power Allocation Scheme for Energy Efficiency Maximization in Distributed Antenna Systems, *IEEE Transactions on Communications*, vol. 63, No. 2, 431–440.

[42] M. El-Aaasser and M. Ashour, (2013) Energy Aware Classification for Wireless Sensor Networks Routing. In: *15th International Conference on Advanced Communication Technology (ICACT)*, pp.66–71, 2013.

[43] F. Etezadi, K. Zarifi, A. Ghrayeb and S. Affes, (2012) Decentralized Relay Selection Schemes in Uniformly Distributed Wireless Sensor Networks, *IEEE Transactions on Wireless Communications*, vol.11, No. 3, 938–951.

[44] D. Dongliang, Q. Fengzhong, Z. Wenshu, and Y. Liuqing, (2011) Optimizing the Battery Energy Efficiency in Wireless Sensor Networks. In: *IEEE International Conference on Signal Processing, Communications and Computing (ICSPCC)*, pp. 1–6, 2011.

[45] G. Brante, R. Demo Souza, and L. Vandendorpe, (2012) Battery-aware Energy Efficiency of Incremental Decode-and-forward with Relay Selection. In: *2012 IEEE Wireless Communications and Networking Conference (WCNC)*, pp. 1108–1112, 2012.

[46] A. G. Marques, W. Xin, and G. B. Giannakis, (2008) Minimizing Transmit Power for Coherent Communications in Wireless Sensor Networks with Finite-Rate Feedback, *IEEE Transactions on Signal Processing*, vol. 56, No. 9, 4446–4457.

[47] C. Wang and S. JuSyue, (2009) An Efficient Relay Selection Protocol for Cooperative Wireless Sensor Networks. In: *IEEE Wireless Communications and Networking Conference, (WCNC 2009)*, pp. 1–5, 2009.

[48] G. Reise and G. Matz, (2012) Optimal Transmit Power Allocation in Wireless Sensor Networks Performing Field Reconstruction. In: *IEEE International Conference on Acoustics, Speech and Signal Processing (ICASSP)*, pp. 3105–3108, 2012.

[49] H. Wan, J.-F. Diouris, and G. Andrieux, (2010) Power Allocation for Virtual MISO Cooperative Communication in Wireless Sensor Networks. In: *Wireless Technology European Conference (EuWIT)*, pp. 69–72, 2010.

[50] L. Juan, H. Jinyu, W. Di, and L. Renfa, (2015) Opportunistic Routing Algorithm for Relay Node Selection in Wireless Sensor Networks, *IEEE Transactions on Industrial Informatics*, vol. 11, No. 1, 112–121.

[51] L. Lingya, H. Cunqing, C. Cailian, and G. Xinping, (2015) Relay Selection for Three-Stage Relaying Scheme in Clustered Wireless Networks, *IEEE Transactions on Vehicular Technology*, vol. 64, No. 6, 2398–2408.

[52] S. Liu, S. Jin, H. Zhu, and K. K. Wong, On Impact of Relay Placement for Energy Efficient Cooperative Networks, *IET Communications*, vol. 8, No. 1, 140–151.

[53] S. Zhengguo, F. Jun, C. H. Liu, V. C. M. Leung, L. Xue, and K. K. Leung, (2015) Energy Efficient Relay Selection for Cooperative Relaying in Wireless Multimedia Networks, *IEEE Transactions on Vehicular Technology*, vol. 64, No. 3, 1156–1170.

[54] G. Reise and G. Matz, (2012) Optimal Transmit Power Allocation in Wireless Sensor Networks Performing Field Reconstruction. *IEEE International Conference on Acoustics, Speech and Signal Processing (ICASSP)*, pp. 3105–3108, 2012.

[55] B. Li, H. Li, W. Wang, Z. Hu, and Q. Yin, (2013) Energy-Effective Relay Selection by Utilizing Spatial Diversity for Random Wireless Sensor Networks, *IEEE Communications Letters*, vol. 17, No. 10, 1972–1975.

[56] P. Li, S. Guo, and J. Hu, J. (2014) Energy-efficient Cooperative Communications for Multimedia Applications in Multi-channel Wireless Networks, *IEEE Transactions on Computers*, vol. 64, No. 6, 1670–1679.

[57] A. M. Akhtar, A. Behnad, and X. Wang, (2014) Cooperative ARQ Based Energy Efficient Routing in Multi-hop Wireless Networks, *IEEE Transactions on Vehicular Technology*, vol. 64, No. 11, 5187–5197.

[58] J. Zhang, D. Zhang, K. Xie, H. Qiao, and S. He, (2017) A VMIMO-based Co-operative Routing Algorithm for Maximizing Network Lifetime, *China Communications*, vol. 14, No. 4, 20–34.

[59] M. El Monser, H. B. Chikha, and R. Attia, (2018) Prolonging the Lifetime of Large-scale Wireless Sensor Networks Using Distributed Cooperative Transmissions, *IET Wireless Sensor Systems*, vol. 8, No. 5, 229–236.

[60] J. Habibi, A. Ghrayeb, and A. Aghdam, (2013) Energy-efficient Cooperative Routing in Wireless Sensor Networks: A Mixed-integer Optimization Framework and Explicit Solution, *IEEE Transactions on Communications*, vol. 61, No. 8, 3424–3437.

[61] C. Y. Aung and P. H. J. Chong, (2017) Cooperative Forwarding in Multi-radio Multi-channel Multi-flow Wireless Networks. In: *IEEE International Conference on Communications*, 21–25 May.

[62] H. Mostafaei, (2019) Energy-efficient Algorithm for Reliable Routing of Wireless Sensor Networks, *IEEE Transactions on Industrial Electronics*, vol. 66, No. 7, 5567–5575.

[63] S. Kim, B-S. Kim, K. H. Kim, and K-I. Kim, (2019) Opportunistic Multipath Routing in Long-hop Wireless Sensor Networks, *Sensors*, vol. 19, No. 19, p.4072.

[64] J. Bai, Y. Sun, C. Phillips, and Y. Cao, (2018) Toward Constructive Relay-based Cooperative Routing in MANETs, *IEEE Systems Journal*, vol. 12, No. 2, 1743–1754.

[65] L. Venkatesh, A. Achar, P. Kushal, and K. R. Venugopal, (2019) Geographic Opportunistic Routing Protocol Based on Two-hop Information for Wireless Sensor Networks, *International Journal of Communication Networks and Distributed Systems*, vol. 23, No. 1, 93–116.

[66] R. Saravanan, (2018) Energy Efficient QoS Routing for Mobile Ad hoc Networks, *International Journal of Communication Networks and Distributed Systems*, vol. 20, No. 3, 372–388.

[67] S. Simi and M. V. Ramesh, (2019) Intelligence in Wireless Network Routing Through Reinforcement Learning, *International Journal of Communication Networks and Distributed Systems*, vol. 23, No. 2, 231–251.

[68] J-H. He, and X-H. Wu, (2007) Variational Iteration Method: New Development and Applications, *Computers and Mathematics with Applications Journal of" Computers, Elsevier*, vol. 54, No. 7, 881–894.

[69] D. P. Kumar, M. S. Babu, and M. S. G. Prasad, (2016) Suboptimal Comparison of AF and DF Relaying for Fixed Target Error Probability. In: *International Conference on Computer Communication and Informatics (ICCCI)*, 7–9 January.

[70] K. Xie, X. Wang, X. Liu, J. Wen, and J. Cao, (2016) Interference-aware Cooperative Communication in Multi-radio Multi-channel Wireless Networks, *IEEE Transactions on Computers*, vol. 65, No. 5, 1528–1542.

[71] M. Wu, D. Wubben, A. Dekorsy, (2011) BER-based Power Allocation for Decode-and-Forward Relaying with M-QAM Constellations. In: *IEEE 7th International Wireless Communications and Mobile Computing Conference*, 4–8 July 2011.

[72] A. M. Akhtar, M. R. Nakhai, (2012) Power Aware Cooperative Routing in Wireless Mesh Networks, *IEEE Communications Letters*, vol. 16, No. 5, 670–673.

[73] J. Gomez-Vilardebo, (2013) Optimal Minimum Energy Routing for Cooperative Multi-hop Wireless Networks. In: *IEEE 24th International Symposium on Personal Indoor and Mobile Radio Communications (PIMRC), 2013.*

[74] X. Li, X. Tao and N. Li, (2016) Energy-Efficient Cooperative MIMO-Based Random Walk Routing for Wireless Sensor Networks, *IEEE Communications Letters*, vol. 20, No. 11, 2280–2283.

[75] D. Praveen Kumar, P. Pardha Saradhi, M. Sushanth Babu, (2017) Shortest Path Cooperative Relay Selection for Multi radio-multi Hop Wireless Networks, *International Journal of Pure and Applied Mathematics*, vol. 117, No. 18, 9–14.

[76] A. M. Akhtar, A. Behnad, and X. Wang, (2015) Cooperative ARQ-Based Energy- Efficient Routing in Multihop Wireless Networks, *IEEE Transactions on Vehicular Technology*, vol. 64, No. 11, 5187–5197.

[77] C. Pandana, W. Pam Siriwongpairat, T. Himsoon, and K. J. Ray Liu, (2006) Distributed Cooperative Routing Algorithms for Maximizing Network Lifetime. In: *IEEE Wireless Communications and Networking Conference, 2006. WCNC 2006.*

[78] L. Zheng, J. Liu, C. Zhai, H. Chen, Y. Zhou, (2011) Energy-efficient Cooperative Routing Algorithm with Truncated Automatic Repeat Request over Nakagami-m Fading Channels, *The Institution of Engineering and Technology Communications*, vol. 5, No. 8, 1073–1082.

[79] B. Rankov and A. Wittneben, (2007) Spectral Efficient Protocols for Half-duplex Fading Relay Channels, *IEEE Journal on Selected Areas in Communications*, vol. 25, No. 2, 379–389.

[80] Y. Han, S. H. Ting, C. K. Ho, and W. H. Chin, (2008) High Rate Two-way Amplify and Forward Half-duplex Relaying with OSTBC. In: *IEEE vehicular technology conference*, pp 2426–2430.

[81] L. Peng, S. Fenggang, Z. Guowei, and H. Jialin (2014) Adaptive Power Allocation for Three Transmission Phases Cognitive Relay Networks with Data Rate Fairness. In: *IEEE international conference on communications and networking in China*, pp 550–553.

[82] A. Naeem and M. H. Rehmani, (2015) Cognitive Relay Networks: A Comprehensive Survey, *EAI Endorsed Transactions on Wireless Spectrum*, vol. 1, No. 3, 1–8.

[83] W. Xu, Y. Wang, J. Lin, and J. Chai, (2017) Outage Performance of Cognitive Radio Networks with Multiple Decode and Forward Relays, *International Journal of Communication System*, vol. 30, No. 8, 1–18.

[84] X. Wang, H. Zhang, T. A. Gulliver, W. Shi, and H. Zhang (2013) Performance Analysis of Two-way AF Cooperative Relay Networks over Weibull Fading Channels, *Journal of Communications*, vol. 8, No. 6, 372–377.

[85] S. Berger, M. Kuhn, A. Wittneben, T. Unger, and A. Klein, (2009) Recent Advances in Amplify-and-forward Two-hop Relaying, *IEEE Communication Magazine*, vol. 47, No. 7, 50–56

[86] T. T. Duy, and H. Y. Kong, (2012) Performance Analysis of Hybrid Decode Amplify- Forward Incremental Relaying Cooperative Diversity Protocol Using SNR-based Relay Selection, *Journal of Communication Network*, vol. 14, No. 6, 703–709.

[87] S. M. Elghamrawy, and A. E. Hassanien, (2018) GWOA: A Hybrid Genetic Whale Optimization Algorithm for Combating Attacks in Cognitive Radio Network, *Journal of Ambient Intelligent Human Computing*, vol. 10, No. 11, 4345–4360.

[88] Y. Li, (2009) Distributed Coding for Cooperative Wireless Networks: An Overview and Recent Advances, *IEEE Communication Magazine*, vol. 47, No. 6, 71–77.

[89] N. Zhao, Y. Chen, R. Liu, M. Wu, and W. Xiong, (2017) Monitoring Strategy for Relay Incentive Mechanism in Cooperative Communication Networks, *Computers & Electrical Engineering*, vol. 60, No. C, 14–29.

[90] M. Kanthimathi, R. Amutha, and K. S. Kumar, (2018) Energy-efficient Differential Cooperative MIMO Algorithm for Wireless Sensor Networks, *Wireless Personal Communications*, vol. 103, No. 4, 2715–2728.

[91] E. Khorov, A. Kiryanov, A. Lyakhov, and G. Bianchi, (2018) A Tutorial on IEEE 802.11 ax High-efficiency WLANs, *IEEE Communications Surveys & Tutorials*, vol. 21, No. 1, 197–216.

[92] T. Lin, K. Wu, and G. Yin, (2015) Channel-hopping Scheme and Channel Diverse Routing in Static Multi-radio Multi-hop Wireless Networks, *IEEE Transactions on Computers*, vol. 64, No. 1, 71–86.

[93] A. Hamzah, M. Shurman, O. Al-Jarrah, and E. Taqieddin(2019) An Energy-efficient Fuzzy-logic-based Clustering Technique for Hierarchical Routing Protocols in Wireless Sensor Networks. *Sensors*, vol. 19, No. 3, 561, 1–23.

[94] M. Ragheb, and S. M. S. Hemami, (2019) Secure Transmission in Large-scale Cooperative Millimeter-wave Systems with Passive Eavesdroppers, *IET Communications*, vol. 14, No. 1, 37–46.

[95] J. Yao, and J. Xu, (2019) Secrecy Transmission in Large-scale UAV-enabled Wireless Networks, *IEEE Transactions on Communications*, vol. 67, No. 11, 7656–7671.

Index

For Product Safety Concerns and Information please contact our EU
representative GPSR@taylorandfrancis.com
Taylor & Francis Verlag GmbH, Kaufingerstraße 24, 80331 München, Germany